CATHETER-RELATED INFECTIONS IN THE CRITICALLY ILL

PERSPECTIVES ON CRITICAL CARE
INFECTIOUS DISEASES
Jordi Rello, M.D., Series Editor

---

1. N. Singh and J.M. Aguado (eds.): Infectious Complications in Transplant Recipients. 2000. ISBN 0-7923-7972-1
2. P.Q. Eichacker and J. Pugin (eds.): Evolving Concepts in Sepsis and Septic Shock. 2001. ISBN 0-7923-7235-2
3. J. Rello and K. Leeper (eds.): Severe Community Acquired Pneumonia. 2001. ISBN 0-7923-7338-3
4. R.G. Wunderink and J. Rello (eds.): Ventilator Associated Pneumonia. 2001. ISBN 0-7923-7444-4
5. R.A. Weinstein and M. Bonten (eds.): Infection Control in the ICU Environment. 2002. ISBN 0-7923-7415-0
6. R.A. Barnes and D.W. Warnock (eds.): Fungal Infection in the Intensive Care Unit. 2002. ISBN 1-4020-7049-7
7. A.R. Hauser and J. Rello (eds.): Severe Infections Caused by Pseudomonas Aeruginosa. 2003. ISBN 1-4020-7421-2
8. N.P. O'Grady and D. Pittet (eds.): Catheter-Related Infections in the Critically Ill. 2004. ISBN 1-4020-8009-3

CATHETER-RELATED INFECTIONS IN THE CRITICALLY ILL

Edited by

Naomi P. O'Grady, M.D.
Critical Care Medicine Department
Warren Grant Magnuson Clinical Center
National Institutes of Health
10 Center Drive
Building 10, Room 7D43
Bethesda, MD 20892

and

Didier Pittet, M.D.
University Hospital
Infection Control Program
Department of Internal Medicine
University of Geneva Hospitals
1211 Geneva 14
Switzerland

Springer Science+Business Media, LLC

Electronic Services <http://www.wkap.nl>

Library of Congress Cataloging-in-Publication Data

Catheter-related infections in the critically ill / edited by Naomi P. O'Grady and Didier Pittet.
 p. ; cm. – (Perspectives on critical care infectious diseases ; 8)
 Includes bibliographical references and index.
  ISBN 978-1-4757-7955-4   ISBN 978-1-4020-8010-4 (eBook)
  DOI 10.1007/978-1-4020-8010-4

 1. Nosocomial infections. 2. Catheterization—Complications. 3. Critical care medicine.
I. O'Grady, Naomi P. II. Pittet, Didier, 1957 – III. Series
 [DLM: 1. Catheterization—adverse effects. 2. Cross Infection—etiology. 3. Critical
 Care—methods. 4. Critical Illness—therapy. 5. Infection Control. WB 365 C363 2004]
RC112.C385 2004
616.02'8—dc22   2004042118

CONTENTS

CONTENTS   v

Contributors   vii

Preface
Naomi P. O'Grady and Didier Pittet   xi

**Perspectives on Critical Care Infectious Disease: An
Introduction to the Series**
Jordi Rello, M.D.   xiii

1. **The Epidemiology of Catheter-Related
   Infection in the Critically Ill**   1
   Nasia Safdar, M.D., Leonard A. Mermel, D.O., Sc.M.
   Dennis G. Maki, M.D.

2. **Epidemiology and Pathogenesis of Catheter-Related
   Bloodstream Infections**   23
   Antonio Sitges-Serra, F.R.C.S. (Ed.)

3. **Diagnosis**   41
   Stephen O. Heard, M.D., F.C.C.M.

4. **Diagnosis of Catheter-Related Infections**   59
   Gérard Nitenberg, M.D., François Blot, M.D.

5. **The Impact of Catheter-Related Infection in
   the Critically Ill**   77
   Christian Brun-Buisson, M.D.

6. **The Impact of Catheter-Related Bloodstream
   Infections**   87
   Karin E. Byers, M.D., M.S., Barry M. Farr, M.D., M.Sc.

7. **Management and Treatment**   99
   Amar Safdar, M.D. and Issam I. Raad, M.D.

8.   **The Management and Treatment of Intravascular**   113
     Catheter-Related Infections
     Professor T.S.J. Elliott

9.   **Education as the Primary Tool for Prevention**   127
     Phillippe Eggimann, M.D., Didier Pittet, M.D., M.S.

10. **Education as an Intervention for Reducing**
     Vascular Catheter Infections   139
     Robert J. Sherertz, M.D.

11. **ICU Prevention Strategies**   147
     Jean-François Timsit, M.D.

12. **Novel Strategies of Preventing Catheter-Related**   159
     Infections in the ICU
     Naomi P. O'Grady, M.D.

Index   173

CONTRIBUTORS

François Blot, M.D.
Service de Réanimation Polyvalente
Institut Gustave Roussy
Villejuif, France

Christian Brun-Buisson, M.D.
Department of Intensive Care and Infection Control Unit
Centre Hospitalier Universitaire Henri Mondor
Assistance Publique Hôpitaux do Paris and Université Paris
Paris, France

Karin E. Byers, M.D., M.S.
University of Pittsburgh Medical Center
Pittsburgh, PA

Phillippe Eggimann, M.D..
Medical Intensive Care Unit and Infection Control Program
Department of Internal Medicine
University of Geneva Hospitals
Geneva, Switzerland

Prof. T.S.J. Elliott
Department of Clinical Microbiology
University Hospital Birmingham NHS Trust
The Queen Elizabeth Hospital, Edgbaston
Birmingham, United Kingdom

Barry M. Farr, M.D., M.Sc.
University of Virginia Health System
Charlottsville, Virginia

Stephen O. Heard, M.D., F.C.C.M.
Department of Anesthesiology
University of Massachussetts Medical Center
Worcester, Massachusetts

Dennis G. Maki, M.D.
Section of Infectious Diseases
Department of Medicine
University of Wisconsin Medical School
Madison, Wisconsin

Leonard A. Mermel, D.O., Sc.M.
Division of Infectious Disease
The Rhode Island Hospital, Brown Medical School
Providence, Rhode Island

Gérard Nitenberg, M.D.
Service de Réanimation Polyvalente
Institut Gustave Roussy
Villejuif, France

Naomi P. O'Grady, M.D.
Warren Magnusen Clinical Center
Critical Care Medicine Department
National Institutes of Health
Bethesda, Maryland

Didier Pittet, M.D., M.S.
Medical Intensive Care Unit and Infection Control Program
Department of Internal Medicine
University of Geneva Hospitals
Geneva, Switzerland

Issam I. Raad, M.D.
The University of Texas
M.D. Anderson Cancer Center
Houston, Texas

Amar Safdar, M.D.
The University of Texas
M.D. Anderson Cancer Center
Houston, Texas

Nasia Safdar, M.D.,
Section of Infectious Diseases
Department of Medicine
University of Wisconsin Medical School
Madison, Wisconsin

Robert J. Sherertz, M.D.
Division of Infectious Diseases
Wake Forest University School of Medicine
Winston Salem
North Carolina

Antonio Sitges-Serra, F.R.C.S (Ed.)
Department of Surgery
Hospital Universitari del Mar
Barcelona, Spain

Jean-François Timsit, Ph..D.
Réanimation Médicale et Infectieuse
Hôpital Bichat-Claude Bernard
Paris, France

Preface

Intravascular catheters are an integral part of the daily practice of medicine in the intensive care unit. As such, management of these catheters poses significant challenges to the practitioner. Vascular access is necessary in the intensive care setting, yet the devices themselves put patients at significant risk for infection. As hospital infection rates are increasingly used as a surrogate marker for measuring patient safety and quality healthcare, preventing catheter-related infection takes on added importance.

It is the intent of this issue to provide the intensivist with a collection of reviews that detail a practical approach to the prevention and management of catheter-related infections and to highlight some of the recent advances in novel technologies and strategies to prevent infection. As patients require catheters for longer periods of time, the types of catheters that are being placed are changing. Although tunneled catheters are still frequently placed in patients who are known to require extended vascular access, peripherally inserted central catheters are rapidly becoming a reasonable alternative, both in the outpatient and the intensive care setting. When temporary central venous catheters are placed, often antibiotic or antiseptic-coated devices are used. Although they are more expensive to purchase, data supports an overall decrease in hospital cost when the price of extra hospital days for infection is factored into the equation. Last but not least, in some institutions, strategies based on educational interventions of critical care staff proved to be extremely efficacious at reducing infection rates and at least as cost-effective as the use of antimicrobial-coated devices, and with no fear about resistance acquisition.

Given the changing types of catheters placed today, the epidemiology and pathogenesis of infection has also changed over the past ten years. Data regarding biofilms and the role they play in catheter infections take on added significance when coupled with the extended dwell times of catheters.

Finally, managing catheter infections in the intensive care unit has become increasingly challenging. Often it is not so simple to take out one catheter and replace it with a new one. Patients often have limited access, coagulopathies, or other anatomical and clinical considerations that preclude removing the central catheter. There are data to support leaving a catheter in place and treating through an infection in certain situations.

We hope that this volume will provide the reader with insights into some of the most interesting and useful data in the field of catheter-related infection. We hope too that the strategies highlighted to prevent infection will be implemented and will have a measurable impact in decreasing rates of infection in the intensive care unit setting. We are very grateful to each of the contributors for the time and effort they have expended to make this a useful and exciting reference tool. We also appreciate Dr. Jordi Rello for the opportunity to prepare this volume in the Perspectives Series. Lastly, we would not have been able to complete this task without the exceptional editorial assistance from Ms. Jennifer Candotti, to whom we are most appreciative.

Naomi P. O'Grady, M.D.
Didier Pittet, M.D., M.S.
Guest Editors

PERSPECTIVES ON CRITICAL CARE INFECTIOUS DISEASES

An Introduction to the Series

Different models of intensive care medicine have been developed worldwide, involving surgeons, anesthetists, internists and critical care physicians. All intensive care departments of hospitals have in common, the highest incidence of antibiotic consumption, the highest incidence of nosocomial infections and are grouping community-acquired infections with high degrees of severity. Intensive care areas of hospitals have the largest number of infection outbreaks and require differentiated strategies of prevention.

The specific characteristics of the involved population require differentiated approaches in diagnosis and therapy from those required in classical infectious problems. The specific pharmacodynamic conditions of patients requiring mechanical ventilation or continuous renal replacement, require participation of experts in pharmacology.

The specific objective of this Series is to update therapeutic implications and discuss controversial topics in specific infectious problems involving critically ill patients. Each topic will be discussed by two authors representing the different management perspectives for these controversial and evolving topics. The Guest Editors, one from North America and one from Europe, have invited contributors to present the most recent findings and the specific infectious disease problems and management techniques for critically ill patients, from their perspective.

Jordi Rello, M.D.
Series Editor

Chapter 1

THE EPIDEMIOLOGY OF CATHETER-RELATED INFECTION IN THE CRITICALLY ILL

Nasia Safdar, M.D., Leonard A. Mermel, D.O., Sc.M.,
Dennis G. Maki, M.D.
*Section of Infectious Diseases (NS, DGM), Department of Medicine, University of Wisconsin
Medical School, Madison, Wisconsin, and the Division of Infectious Disease (LAM), the
Rhode Island Hospital, Brown Medical School, Providence, Rhode Island*

Introduction

Vascular access is one of the most essential features of modern critical care medicine. In the Intensive Care Unit (ICU), the entire range of intravascular devices (IVDs) needed for vascular access is encountered: central venous catheters of every type, including noncuffed multilumen catheters, large dual-lumen catheters for hemodialysis, large introducers and flow-directed, balloon-tipped pulmonary artery (Swan-Ganz) catheters, cuffed and tunneled Hickman-like CVCs, arterial catheters used for hemodynamic monitoring, small peripheral venous catheters and, increasingly peripherally-inserted central venous catheters (PICCs).

Unfortunately, the IVDs needed to establish reliable access are associated with significant potential for producing iatrogenic disease, particularly bacteremia and candidemia (1-3), deriving from infection of the percutaneous device used for vascular access or from contamination of the infusate administered through the device (4).

The forms of infection associated with IVDs range from exit site infection -- purulence, inflammation and erythema at the site; local infection, usually asymptomatic, synonymous with colonization of the catheter; bloodstream infection (BSI) , the most serious, potentially fatal complication of IVDs and the gravest infectious complication of vascular access, septic thrombophlebitis of peripheral veins and septic thrombosis of the great central veins (Table 1) (5).

MAGNITUDE OF THE PROBLEM

More than 250,000 IVD-related BSIs occur in the United States each year (1); the majority are related to short-term noncuffed, percutaneously-placed central venous catheters (CVCs). Whereas earlier studies have found a 12-25% attributable mortality of IVD-related BSI (6-9), recent studies have questioned the attributable mortality of IVD-related BSIs and primary BSI; 10-13 however, these infections are associated with prolongation of hospital stay (7-9, 14) and marginal cost to the health system of $33,000 to 35,000 per episode (7, 8, 14) The risk is greatly amplified in the ICU setting where at least 80,000 IVDR BSIs occur annually (3,15) with a marginal cost of $33,000 to $71,000 per case (15).

The magnitude of risk of IVD-related BSI varies with the type of IVD in place (Table 2) (16). The device that poses the greatest risk of IVDR BSI today is the CVC in its many forms: short-term noncuffed, single- or multi-lumen catheters inserted percutaneously into the subclavian, internal jugular or femoral vein have been associated with rates of catheter-related BSI in the range of 3 to 5% (2- 3 per 1000 IVD-days). Far lower rates of infection occur with surgically implanted cuffed Hickman or Broviac and subcutaneous central venous ports (1 and 0.2 per 1000 IVD-days, respectively). Contrary to popular belief, PICCs used in a hospitalized population, and arterial catheters are associated with a risk of catheter-related BSI approaching that seen with noncuffed multilumen CVCs; up to 2.1% and 3.7 BSIs per 1000 IVD-days, and 1.5% and 2.9 per 1000 IVD-days, respectively. The increased risk observed with PICCs in hospitalized patients is especially of importance as PICC sales in the U.S have risen greatly and are expected to continue to rise.

Rates of IVDR-BSI are also influenced by severity of illness and underlying diseases: granulocytopenic patients, HIV, and those undergoing marrow transplantation have a much higher risk of IVDR BSI. However, risk can be greatly reduced by good catheter care practices and consistent application of strategies shown to reduce risk of IVDR-related BSI.

Table 1. The Spectrum Of Catheter-Related Infection.

Infection	Definitions
Catheter colonization	Significant growth of a microorganism in a quantitative or semiquantitative culture of the catheter tip, subcutaneous catheter segment or catheter hub
Phlebitis	Induration or erythema, warmth, and pain or tenderness around catheter exit site
Exit-site infection	
Microbiological	Exudate at catheter exit-site yields a microorganism with or without concomitant bloodstream infection may also occur
Clinical	Erythema, induration and/or tenderness within 2 cm of the catheter exit site; may be associated with other signs and symptoms of infection, such as fever, or pus emerging from the exit site, with or without concomitant bloodstream infection
Tunnel infection	Tenderness, erythema or site induration > 2 cm from the catheter site along the subcutaneous tract of a tunneled (e.g. Hickman or Broviac) catheter, in the absence of concomitant bloodstream infection
Pocket infection	Infected fluid in the subcutaneous pocket of a totally implanted intravascular device, often associated with tenderness, erythema, and/or induration over the pocket; spontaneous rupture and drainage, or necrosis of the overlying skin , with or without concomitant bloodstream infection may occur
Bloodstream infection	
Infusate related	Concordant growth of the organism from infusate and cultures of percutanouesly obtained blood samples with no other identifiable source of sepsis
Catheter-related	Bacteremia or fungemia in a patient who has an intravascular device and > or 1 positive result of culture of blood samples obtained from peripheral vein, and no apparent source for bloodstream infection (with the exception of the catheter). One of the following should be present: a positive result of semiquantitiative (>= 15 CFU per catheter segment) or quantitiative (>= 100 cfu per catheter segment) catheter culture, whereby the same organism (species and antibiogram) is isolated from a

catheter segment and a peripheral blood sample;
simultaneous quantitative cultures of blood samples,
with a ratio of >=5: 1 (CVC vs peripheral) ;differential
time to positivity 2 hours earlier than positive result
from peripheral blood.

Septic
thromboplebitis/thrmobosis

  Peripheral vein   Palpable cord (thrombosed vessel), with expression of
                                       pus from exit site, or aspiration of thrombosed
                                       peripheral vein showing organisms.
  Central vein   Overwhelming sepsis with high grade bacteremia or
                                       candidemia. With complete occlusion of the subclavian
                                       vein, the ipsilateral arm may be swollen; if there is
                                       involvement of the internal jugular vein, the face and
                                       neck may swell.

Adapted from the Infectious Diseases Society of America Guideline for management of
IVD-related Infection (5). Abbreviations: BSI – bloodstream infection; CVC – central
venous catheter, IV – intravenous, IVD – intravascular device

Table 2. Rates Of Bloodstream Infection Caused By Various Types Of Devices Used For Vascular Access*.

Catheter Type	Studies, n	Catheters, n	Mean Duration (Days)	Rates of IVD**-related BSI			
				Per 100 Catheters		Per 1000 IVD-days	
				Pooled Mean	95% CI	Pooled Mean	95% CI
Peripheral venous cannulas							
Plastic catheters	13	5,658	2.6	0.2	0.1 - 0.3	0.6	0.3 - 1.2
Steel needles	2	239	2.6	0.4	0.0 - 2.3	1.6	0.0 - 9.1
Cutdowns	1	27	4.2	3.7	0.1 - 20.6	8.8	0.2 - 49.3
Arterial catheters used for hemodynamic monitoring	6	1,295	5.3	1.5	0.9 - 2.4	2.9	1.8 - 4.5
Central venous catheters							
Standard noncuffed, nonmedicated	61	15,315	14.8	3.3	3.3 - 4.0	2.3	2.0 - 2.4
Antibiotic-coated	3	563	7.3	0.2	0.0 - 1.0	0.2	0.1 - 1.4
Control nonmedicated§	2	184	5.8	4.4	3.5 - 5.5	---	3.4 - 5.4
Antiseptic-impregnated	14	2.081	10.5	3.1	2.4 - 3.9	2.9	2.3 - 3.7
Control nonmedicated§	13	1,717	10.3	3.8	1.5 - 7.8	---	2.6 - 13.5
Peripherally-inserted (PICCs)	8	775	30.0	1.2	0.5 - 2.2	0.4	0.2 - 0.7
Silver-impregnated cuffs	6	481	10.3	3.3	1.9 - 5.4	3.2	1.9 - 5.3
Tunneled but noncuffed	6	693	69.0	12.4	10.7 - 16.6	1.8	1.4 - 2.0
Pulmonary-artery catheters	12	1.014	3.4	1.9	1.1 - 2.5	5.5	3.2 - 12.4
Heparin-bonded	3	1.814	1.7	0.4	0.2 - 0.9	2.6	1.1 - 5.2
Hemodialysis catheters							
Noncuffed	15	997	56.8	16.2	13.5 - 18.3	2.8	2.3 - 3.1
Cuffed	5	427	56.0	6.3	4.2 - 9.2	1.1	0.7 - 1.6
Tunneled and cuffed CVCs	18	2.220	179.6	20.9	18.2 - 21.9	1.2	1.0 - 1.3
Subcutaneous central ports	13	1,582	288.6	5.1	4.0 - 6.3	0.2	0.1 - 0.2
Peripherally-inserted central ports	3	130	199.8	0.0	0.0 - 2.8	0.0	0.0 - 0.1

From Kluger *et al.* Programs and Proceedings of the Fourth Decennial International Conference on Nosocomial and Healthcare-associated Infections, 2000 [Abstract] (16).

* Based on pooled data from 206 published, prospective clinical studies in which every device was evaluated for infection.

** IVD = Intravascular device

§ The comparator catheters in the randomized comparative trials.

PATHOGENESIS

There are two major sources of IVD-related BSI: 1) colonization of the IVD, catheter-related infection, and 2) contamination of the fluid administered through the device, infusate-related infection   Contaminated infusate is the cause of most epidemic intravascular device-related BSIs; in contrast, catheter-related infections are responsible for most endemic device-related BSIs.

Understanding the pathogenesis of IVD-related BSIs is essential to devising strategies for prevention of these infections; however few published studies have determined the mechanism of IVD-related colonization and infection using sophisticated molecular techniques to prove or disprove potential routes of infection.

In order for microorganisms to cause catheter-related infection they must first gain access to the extraluminal or intraluminal surface of the device where they can adhere and become incorporated into a biofilm that allows sustained colonization and, ultimately hematogenous dissemination   Microorganisms gain access to the implanted IVD by one of three mechanisms: skin organisms invade the percutaneous tract, probably facilitated by capillary action, at the time of insertion or in the days following; microorganisms contaminate the catheter hub (and lumen) when the catheter is inserted over a percutaneous guidewire or later manipulated; or organisms are carried hematogenously to the implanted IVD from remote sources of local infection, such as a pneumonia (Figure 1).

With short-term IVDs (in place <10 days) -- peripheral IV catheters, arterial catheters and non-cuffed, non-tunneled CVCs -- most catheter-related BSIs are of cutaneous origin, from the insertion site, and gain access extraluminally, occasionally intraluminally   For long-term catheters-- tunneled, cuffed CVCs, totally implantable ports and PICCs-- luminal colonization has been shown to be the major mechanism leading to BSI (17,18).   A characteristic pulsed-field gel electrophoresis image obtained from a short-term noncuffed CVC causing BSI is shown in Figure 2 and from a long-term cathter (PICC), in Figure 3.

MICROBIOLOGY

The distribution of microorganisms that cause IVD-related BSIs vary by the type of device used (Table 3) (19). For example, microorganisms found on patient's skin, which gain access to the IVD extraluminally, occasionally, intraluminally—coagulase-negative staphylococci (39%), *Staphylococcus aureus* (26%), and Candida spp. (11%)—account for 76% of IVD-related BSIs with short-term noncuffed devices of all types; only 14% are caused by gram-negative bacilli. In contrast, with long-term surgically implanted devices, such as cuffed and tunneled catheters, PICCs, and subcutaneous central venous ports, gram-negative bacilli, which gain access intraluminally and grow rapidly within the infusate in the device, account for nearly half of IVD-related BSIs; only 2% are caused by Candida spp.

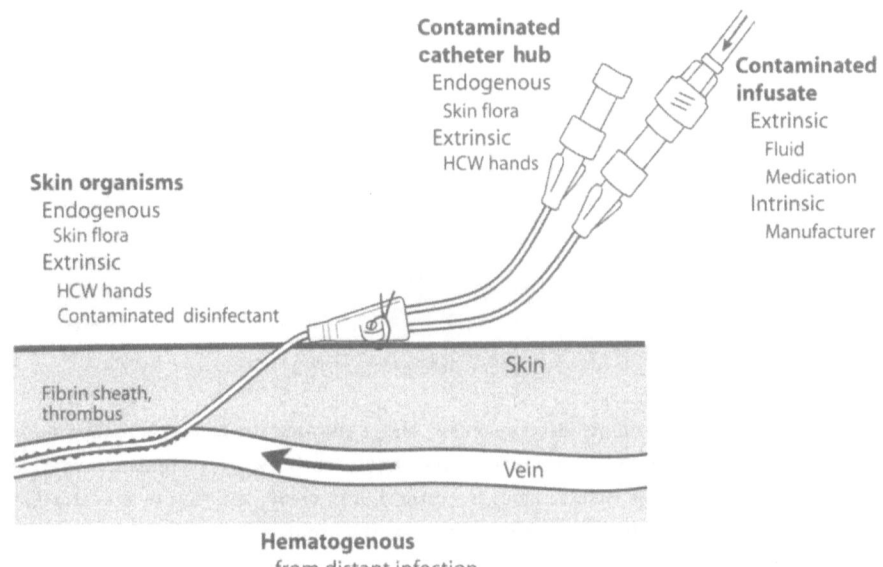

Figure 1. Potential sources of infection of a percutaneous IVD: the contiguous skin flora, contamination of the catheter hub and lumen, contamination of infusate, and hematogenous colonization of the IVD from distant, unrelated sites of infection (from C.J Crnich and D.G Maki (4)).

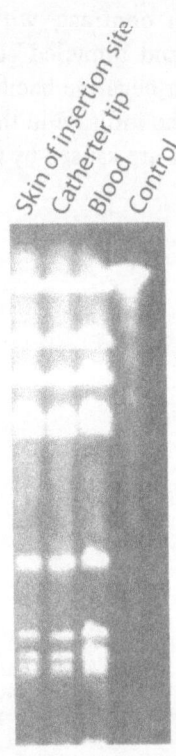

Figure 2. Pulsed-field gel electrophoresis image showing the probable pathogenesis of a central venous catheter-related bacteremia with coagulase-negative Staphylococcus. The isolates from the catheter tip, blood, and skin of the insertion site were all concordant, indicating an extraluminal route of infection (29).

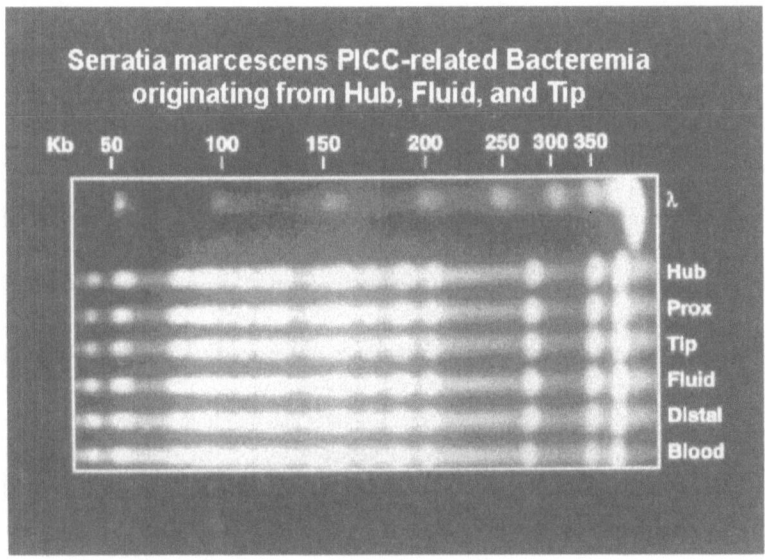

Figure 3. Pulsed-field gel electrophoresis image showing the probable pathogenesis of a PICC-related bacteremia with Serratia marcescens. The isolates from the catheter tip, blood, hub and fluid were all concordant, indicating an intraluminal route of 7infection.

Table 3. Microbial profile of IVD-related infection with percutaneously-inserted short-term noncuffed IVDs and with surgically-implanted long-term central venous devices.

Type of IVD	No. Studies	No. IVDs	CNS	*Staph. aureus*	Ecoc	Enteric GNB	*Pseudomonas aeruginosa*	Candida	Other
						Microorganisms, % of Total			
Percutaneously-inserted noncuffed catheters*									
Colonized	87	3,640	66	6	2	8	4	6	8
IVDR BSI	83	592	40	26	4	10	5	11	8
Surgically-implanted central venous IVDs†									
Local infection	14	117	31	34	0	7	15	2	12
IVDR BSI	38	865	25	13	4	22	9	3	9

From Kluger *et al*. Programs and Proceedings of the 40th Annual Meeting of the Infectious Diseases Society of America , 2002 [Abstract].

Abbreviations: IVD, intravascular device; CNS, coagulase-negative staphylococci; GNB, gram-negative bacilli;

Ecoc, Enterococcus spp., Other, miscellaneous species; IVDR BSI, IVD-related bloodstream infection.

* Peripheral IV catheters, arterial catheters, single- and multilumen CVCs, PA catheters, coated CVCs and noncuffed hemodialysis catheters.

† Cuffed hemodialysis catheters, PICCs, cuffed and tunneled Hickman-like catheters and central venous ports.

RISK FACTORS FOR INFECTION AND IMPLICATIONS FOR PREVENTION

IVDR-BSIs are largely preventable. Strategies for prevention can be successful only if based upon a sound understanding of the risk factors and pathogenesis of IVD-related BSI. A growing body of literature in recent years has greatly enhanced our understanding of the risk factors for IVD-related BSI; a recent review summarizes the major risk factors with short-term noncuffed CVCs (Tables 4,5, and 6) (20).

Table 4. Risk Factors for Catheter-Related BSI with Non-Cuffed Percutaneously-Inserted CVCs.

Risk Factors (No. of studies)	Risk*
Patient Characteristics	
Age (3)	1.0
Female sex (6)	0.1-10
Underlying disease:	
AIDS (2)	4.8
Low CD4 (1)	3.45
Neutropenia (2)	1.0-15.1
GI disease (1)	2.4
Surgical service (1)	4.4
ICU/CCU placement (3)	0.4-6.7
Extended hospitalization (3)	1.0-6.7
Coexistence of other intravascular devices (2)	1.0-3.8
Systemic antibiotics (3)	0.1-0.45
Active infection at any other site (2)	8.7-9.2
Low birth weight (2)	5.13-9.1
High APACHE III score (1)	4.19
Mechanical ventilation (1)	1.97-2.5
Transplantation (1)	2.6
Features of Insertion Site	
Insertion of by house-staff or student (1)	1.0
Difficult insertion (1)	5.4
Maximal sterile barriers (1)	0.2
Tunnelling a non-cuffed CVC (2)	0.3-1.0
Insertion in an old site over a guidewire (8)	1.0-3.3
Insertion site:	
Internal jugular vein (6)	1.0-3.3
Subclavian vein (5)	0.4-1.0
Femoral vein (2)	3.3-4.83
Defatting insertion site (1)	1.0
Cutaneous antiseptic used:	
Chlorhexidine vs. povidone iodine (2)	0.2-0.9
Topical anti-infective cream:	
Povidone-iodine (1)	1.1
Mupirocin (1)	0.3

Risk Factors (No. of Studies) (con't)	Risk*
Chlorhexidine-impregnated dressing (3)	0.3-1.2
Type of dressing (transparent vs. gauze) (6)	0.7-2.8
Colonization of insertion site (4)	6.3-56.5
Catheter characteristics	
Multi-lumen vs. single-lumen catheter (8)	1.0-6.5
Ant-infective coating	
Antibacterial (2)	0.1-0.3
Antiseptic (10)	0.2-1.0
Antibiotic vs. Antiseptic (2)	0.1-1.0
Silver-impregnated cuff (6)	0.3-1.0
Contamination-resistant hub (1)	0.2
Subsequent catheter management	
Routine change of IV set (2)	1.0
Staffing in SICU (Nurse: patient ratio) (1):	
1:2.0	61.5
1:1.5	15.6
1:1.2	4.0
1:1	1.0
Inappropriate catheter usage (1)	5.3
Duration of catheter >6 days (5)	1.0-8.7
Colonization of catheter hub (3)	17.9-44.1
Disinfection of catheter hub (1)	1.2
Blood sampling (1)	1.4
Heparinization (1)	0.9
Parenteral nutrition (2)	1.04-4.79

From Safdar N et al. Medicine 2002 with permission

Abbreviations: CVC, central venous catheter; AIDS, Acquired immunodeficiency syndrome; GI=gastrointestinal tract; ICU, intensive care unit; CCU, critical care unit; SICU, surgical intensive care unit.

* Relative risk or odds ratio

Table 5. Risk Factors for Catheter-Related BSI with Non-Cuffed Pulmonary-Artery Catheters.

Risk Factor (No. of Studies)	Risk*
Insertion with less stringent barrier precautions (1)	2.1
Insertion site in internal jugular	1.0-4.3
Insertion site colonization >100 CFU (1)	5.5
Duration of catheterization >72 hours (3)	1.0-14.4

Table 6. Risk Factors for Catheter-Related BSI with Peripheral Arterial Catheters.

Risk Factor (No. of Studies)	Risk*
Increasing duration of catheterization	5.0-10.0
Insertion using surgical cutdown	14.0
Local inflammation	10.0

Training and Experience of the Inserter

CVCs are associated with significant potential for life-threatening iatrogenic complications besides catheter-related BSI, including pneumothorax, vascular injury, arrhythmias and thromboembolism. Armstrong *et al.* identified inserter experience as an important risk factor for CVC-related BSI in a prospective study of 169 catheters (21). Moreover, a survey of U.S. academic medical centers has shown that up to one-half of clinicians who use PA catheters have major gaps in their understanding of when to use the catheter and how to interpret the data derived from it (22). Only in recent years are U.S. institutions requiring formal training of house officers in the techniques of vascular access. Intensified training and educational programs can greatly reduce the baseline risk of CVC-related BSI in a center.

Intravenous Therapy Teams

Good technique is essential. Studies have shown that the use of special IV therapy teams, consisting of trained nurses or technicians who can assure a consistent and high level of aseptic technique during catheter insertion and in follow-up care of the catheter, have been associated with substantially lower rates of catheter-related BSI and are cost-effective.

Institutional IV teams should be encouraged, but even if an institution does not have an IV team, it can greatly reduce its rate of IVDR BSI by formal education of nurses and physicians and strict adherence to IVD care protocols (23).

Sterile Barrier Precautions

Mermel *et al.* (24) found in a prospective study of 302 pulmonary-artery catheters that failure to use maximal sterile barriers at the time of catheter insertion increased the risk of catheter-related infection more than two-fold (RR, 2.1). Whereas the issue has not been studied extensively, in one well-controlled randomized trial it was found that the use of maximal sterile barriers when inserting a CVC in a patient with cancer greatly reduced the risk of CVC-related BSI (RR, 0.20) (25).
It seems clear that physicians inserting a CVC should wear a long-sleeved sterile surgical gown and sterile gloves and, to be in compliance with universal precautions, a mask and eye cover; the potential insertion site should be draped with a large sterile sheet (23). Maximal sterile barrier precautions are not necesssary for peripheral arterial catheters used for hemodynamic monitoring, where sterile gloves and a sterile fenestrated drape will suffice based on a a prospective study showing no difference in colonization but the study was underpowered to show a difference in catheter-related BSI (26).

Site of Insertion

At least six studies, including one randomized clinical trial, have found that percutaneous insertion of a CVC in an internal jugular or femoral vein is associated with a substantially higher risk of catheter-related BSI than insertion in a subclavian vein (RR, 1-3.3) (24,27-31). Femoral line insertion also dramatically and independently increases the risk of the life-threatening complications deep venous thrombosis (30). Whereas placement in an internal jugular or femoral vein is associated with less risk of pneumothorax and permits control of local hemorrhage by the application of pressure, the risk of mechanical complications with central venous cannulation, such as pneumothorax or hemorrhage, has greatly declined in recent years (59), reflecting better training in the techniques of percutaneous catheter insertion and greater experience. It should be possible to place a CVC percutaneously in the subclavian vein with a very low risk of barotrauma, in the range of 1% or less.

We believe these data indicate that training programs should strive to encourage use of the subclavian vein as the preferred site of access for CVCs (23) (other than catheters needed for long-term hemodialysis), and should assure that house officers are trained in establishing central access in the subclavian vein. Catheterization of the femoral vein should be kept to a strict minimum and if accessed during a code situation, the catheter should be changed to an alternative site as soon as it's safe to do so. Tunneling a CVC appears to reduce the risk of catheter-related BSI, both with catheters placed in the internal jugular or femoral veins, and might be considered if circumstances mandate cannulation of an internal jugular or femoral vein rather than a subclavian vein (e.g., severe coagulopathy or a hemodialysis catheter).

Catheter Exchange Over a Guidewire

The Seldinger technique for catheter insertion has been a major advance, permitting the great central veins to be cannulated with considerably less risk of pneumothorax and vascular injury. To avoid iatrogenic mechanical complications associated with percutaneous insertion of another CVC, new catheters are commonly inserted over a guidewire in the site of an old catheter. Numerous studies have examined the impact of this practice on the risk of infection (32-43), most did not utilize multivariable techniques. Eight randomized trials to address this issue have had conflicting results (33-37,42-44). The best prospective randomized trial, which included pulmonary-artery catheters, found a nearly two-fold increased risk of catheter-related BSI with CVCs replaced on a periodic basis in old sites over a guidewire (9 vs 5 cases per 1000 catheter-days); 75 percent of all catheter-related BSIs in the study population occurred within 72 hours of catheter exchange over a guidewire (35). However, a systematic review of the effect of guidewire exchange and new site replacement strategies for central venous catheters in critically ill patients did not find a statistically significant reduction in catheter-related BSI with routine guidewire exchange (RR 1.72, 95% CI 0.12-1.91) (45).

If a CVC is replaced because of suspected infection without signs of sepsis, or the catheter has malfunctioned (e.g it is cracked), it is reasonable to replace a catheter in the same site over a guidewire if the patient has limited sites for new access or would be at a very high risk for percutaneous central venous cannulation in a new site (e.g., coagulopathy or morbid obesity) (23). However, it is imperative that the same meticulous aseptic technique and use of full sterile barriers that are mandatory during the insertion of any new CVC be employed. After vigorously cleansing the site with the antiseptic solution, inserting the guidewire, removing the old catheter and cleansing the site once more with the antiseptic solution, the operator should reglove and ideally redrape the site, as the original gloves and drapes are likely to have become contaminated from manipulation of the old catheter.

It is also essential to routinely culture the old catheter and, if the patient is febrile or shows other signs of sepsis, to obtain blood cultures (23). If these cultures demonstrate that the old catheter was infected, the new catheter placed in an old site should ideally be immediately removed to prevent progression to catheter-related BSI or perpetuation of ongoing BSI, as a new catheter has been inserted into an infected tract; need for continued access would mandate placement of a new catheter in a new site. If culture of the old catheter shows that it is not colonized, it has been possible to preserve access and exclude it as the cause of fever and sepsis without subjecting the patient to the hazards associated with percutaneous insertion of a new catheter.

In general, if an old insertion site is inflamed, especially if there is purulence, the patient shows signs of sepsis that might be originating from the catheter or the patient has cryptogenic bacteremia or candidemia, it is strongly recommended that a new catheter not be inserted over a guidewire into an old, potentially-infected site (23).

HEAVY COLONIZATION OF THE INSERTION SITE AND CUTANEOUS ANTISEPSIS

Colonization of the insertion site will be greatly influenced by the choice of the site for insertion. In a prospective study, it was found that the density of the transient cutaneous microflora was highest at the base of the neck, the site of insertion of an internal jugular vein catheter, as contrasted with over the upper chest, the site for insertion of a subclavian vein catheter. In neonates, there is a significantly greater density of microbes in the combined jugular and femoral sites than either the umbilical or subclavian site.

Given the powerful evidence for the importance of cutaneous micro-organisms and particularly the density of the microflora at the potential insertion site in the pathogenesis of CVC-related infection, measures to reduce cutaneous colonization of the insertion site would seem of the highest priority, particularly the choice of the chemical antiseptic used for disinfecting the site. In the United States, iodine-based disinfectants, particularly iodophors such as 10% povidone-iodine, are used most widely. Chlorhexidine, a biguanide with potent and broad-spectrum activity, exhibits prolonged antimicrobial activity on the skin surface after a single application, in contrast to alcohol or iodine-based antiseptics. To date, seven prospective randomized clinical trials have compared the efficacy of 10% povidone-iodine and chlorhexidine antisepsis for vascular access. The largest, a prospective randomized trial with 750 CVCs and arterial catheters used in patients in an ICU, showed that 2% chlorhexidine was superior to 10% povidone-iodine or 70% alcohol for prevention of CVC-related BSI (RR, 0.16). In six of the seven trials to date, chlorhexidine was superior to povidone-iodine for preventing catheter colonization, and in two, CVC-related sepsis was reduced significantly.

These studies in aggregate indicate that a 0.5 - 2% chlorhexidine-alcohol tincture or a 1-2% aqueous solution is more effective than iodophors or 70% alcohol for prevention of CVC-related colonization and BSI. Two recent meta-analyses of randomized trials comparing chlorhexidine to 10% povidone-iodine for cutaneous antisepsis found a 50% reduction in the risk of CVC-related BSI with the use of chlorhexidine.

Disinfection of skin should be done with an appropriate antiseptic prior to catheter insertion and at the time of dressing changes. A 2% chlorhexidine-based preparation is preferred. Alternatively, tincture of iodine, an iodophor, or alcohol could be used. Allow the antiseptic to remain on the insertion site and to dry before inserting the catheter. Allow povidone-iodine to remain on the skin for at least 2 minutes, or longer if it is not yet dry before inserting the catheter.

Site Dressings

The importance of the cutaneous microflora in the pathogenesis of CVC-related infection suggests that the dressing applied to the insertion site could have considerable influence on the incidence of catheter-related infection. In recent years, transparent polyurethane film dressings have become available. They secure the device more reliably, permit continuous inspection of the site, and are generally more comfortable than gauze and tape; moreover, they permit patients to bathe and shower without saturating the dressing. Studies of polyurethane dressings on short-term non-cuffed CVCs have yielded conflicting results; however, a meta-analysis of the largest and most rigorously controlled randomized trials has shown that these dressings do not materially increase the risk of CVC-related BSI (RR 0.99, 95% CI 0.90-1.09) (46).

Either sterile gauze or sterile, transparent, semipermeable dressing may be used to cover the catheter site. If the patient is diaphoretic, or if the site is bleeding or oozing, a gauze dressing is preferable to a transparent, semipermeable dressing (23).

Manipulations of the System

Contamination of infusate, stopcocks or catheter hubs, the cause of many CVC-related BSIs, has been the cause of most outbreaks of infusion-related bacteremia or candidemia.

In general, running infusions should be manipulated as little as possible, and persons handling or entering the system should first wash their hands or don clean gloves (23). Efforts should be made to limit entry into the monitoring circuit for the purpose of drawing blood or other tests (23). The number of stopcocks in the system should also be kept to a minimum. It is unknown whether wiping a stopcock which has been opened with an anti-infective agent might be of value.

Prolonged Catheter Placement

Exactly how long non-cuffed short-term CVCs can be left in place safely, particularly in critically ill patients in an ICU, has not been adequately assessed. In general, however, most studies that have examined duration of placement as a risk factor have shown that prolonged placement significantly increases the cumulative risk of infection, particularly insertions longer than 5-7 days (27,37) A simple but elegant mathematical model has also been derived demonstrating the increased risk of catheter-related bloodstream infection for each day of catheterization (47). The need for continued use of an intravascular catheter should be frequently reassesed and the device should be removed as soon as the intended use is over (48).

Finally, what has not been conclusively established is whether routine replacement of a non-cuffed CVC to a new site at periodic intervals, such as every 4-5 days, significantly reduces the risk of CVC-related BSI in patients requiring prolonged central access. While some studies report no decline in the incidence of CVC-related BSI with routine replacement (35,37,49), most that have examined this issue have not had sufficient statistical power to answer the question (37,44,50). In the absence of conclusive data affirming benefit, central venous or arterial catheters should not be routinely replaced solely for the purpose of reducing the risk of catheter-related infection (23).

The question thus remains unanswered; however, the availability of novel technology may obviate this concern. The studies of anti-infective-coated CVCs show a sufficiently reduced risk of CVC-related BSI that it would appear that with the use of such technology in patients requiring prolonged central access, it should be safe to leave a CVC in place for 10-20 days, if necessary, perhaps even longer if the device is dedicated to total parenteral nutrition or anti-infective therapy. Moreover, the use of chlorhexidine-impregnated dressings or engineered contamination-resistant catheter hubs can also reduce risk and permit prolonged cannulation with a very low risk of infection.

REFERENCES

1.   Maki DG, Mermel LA. Infections due to infusion therapy. In: Bennet JV, Brachman PS, eds. Hospital Infections. Philadelphia: Lippincott-Raven Publishers; 1998:689-724.
2.   Raad I. Intravascular catheter-related infections. Lancet. 1998;351:893-898.
3.   Mermel LA. Prevention of intravascular catheter-related infections. Ann Intern Med. 2000;132:391-402.
4.   Crnich CJ, Maki DG. The promise of novel technology for the prevention of intravascular device-related bloodstream infection. I. Pathogenesis and short-term devices. Clin Infect Dis. 2002;34:1232-1242.

5. Mermel LA, Farr BM, Sherertz RJ, *et al.* Guidelines for the management of intravascular catheter-related infections. J Intraven Nurs. 2001;24:180-205.
6. Smith RL, Meixler SM, Simberkoff MS. Excess mortality in critically ill patients with nosocomial bloodstream infections. Chest. 1991;100:164-167.
7. Pittet D, Tarara D, Wenzel RP. Nosocomial bloodstream infection in critically ill patients:excess length of stay, extra costs and attributable mortality. JAMA. 1994;271:1598-1601.
8. Arnow PM, Quimosing EM, Beach M. Consequences of intravascular catheter sepsis. Clin Infect Dis. 1993;16:778-784.
9. Collignon PJ. Intravascular catheter associated sepsis: a common problem. The Australian Study on Intravascular Catheter Associated Sepsis. Med J Aust. 1994;161:374-378.
10. Soufir L, Timsit JF, Mahe C, Carlet J, Regnier B, Chevret S. Attributable morbidity and mortality of catheter-related septicemia in critically ill patients: a matched, risk-adjusted, cohort study. Infect Control Hosp Epidemiol. 1999;20:396-401.
11. Digiovine B, Chenoweth C, Watts C, Higgins M. The attributable mortality and costs of primary nosocomial bloodstream infections in the intensive care unit. Am J Respir Crit Care Med. 1999;160:976-981.
12. Pelletier SJ, Crabtree TD, Gleason TG, Pruett TL, Sawyer RG. Bacteremia associated with central venous catheter infection is not an independent predictor of outcomes. J Am Coll Surg. 2000;190:671-680; discussion 680-671.
13. Renaud B, Brun-Buisson C. Outcomes of primary and catheter-related bacteremia. A cohort and case-control study in critically ill patients. Am J Respir Crit Care Med. 2001;163:1584-1590.
14. Rello J, Ochagavia A, Sabanes E, *et al.* Evaluation of outcome of intravenous catheter-related infections in critically ill patients. Am J Respir Crit Care Med. 2000;162:1027-1030.
15. Crnich CJ, Maki DG. The role of intravascular devices in sepsis. Curr Infect Dis Rep. 2001;3:497-506.
16. Kluger D, Maki D. The relative risk of intravascular device-related bloodstream infections with different types of intravascular devices in adults. A meta-analysis of 206 published studies. Paper presented at: Programs and Proceedings of the Fourth Decennial International Conference on Nosocomial and Healthcare-Associated Infections, 2000; Atlanta, GA, 2000.
17. Sitges-Serra A, Puig P, Linares J, *et al.* Hub colonization as the initial step in an outbreak of catheter-related sepsis due to coagulase negative staphylococci during parenteral nutrition. JPEN J Parenter Enteral Nutr. 1984;8:668-672.
18. Raad I, Costerton W, Sabharwal U, Sacilowski M, Anaissie E, Bodey GP. Ultrastructural analysis of indwelling vascular catheters: a quantitative relationship between luminal colonization and duration of placement. J Infect Dis. 1993;168:400-407.
19. Maki DG, Kluger DM, Crnich CJ. The microbiology of intravascular device-related (IVDR) infection in adults: an analysis of 159 prospective studies; implications for prevention. Paper presented at: 40th Annual Meeting of the Infectious Diseases Society of America, 2002; Chicago, IL.

20. Safdar N, Kluger DM, Maki DG. A review of risk factors for catheter-related bloodstream infection caused by percutaneously inserted, noncuffed central venous catheters: implications for preventive strategies. Medicine (Baltimore). 2002;81:466-479.

21. Armstrong CW, Mayhall CG, Miller KB, *et al.* Prospective study of catheter replacement and other risk factors for infection of hyperalimentation catheters. Journal Infect Dis. 1986;154:808-816.

22. Iberti TJ, Fischer EP, Leibowitz AB, *et al.* A multicenter study of physicians' knowledge of the pulmonary artery catheter. JAMA. 1990;264:2928-2932.

23. O'Grady NP, Alexander M, Dellinger EP, *et al.* Guidelines for the prevention of intravascular catheter-related infections. Centers for Disease Control and Prevention. MMWR Recomm Rep. 2002;51:1-29.

24. Mermel LA, McCormick RD, Springman SR, Maki DG. The pathogenesis and epidemiology of catheter-related infection with pulmonary artery Swan-Ganz catheters: a prospective study utilizing molecular subtyping. Am J Med. 1991;91(Suppl 3B):1897-1205

25. Raad, II, Hohn DC, Gilbreath BJ, *et al.* Prevention of central venous catheter-related infections by using maximal sterile barrier precautions during insertion. Infect Control . Hosp Epidemiol. 1994;15:231-238.

26. Rijnders BJ, Van Wijngaerden E, Peetermans WA. Use of full sterile barrier precautions during insertion of arterial catheters: a randomized trial. Clin Infect Dis. 2003;36:743-748.

27. Richet H, Hubert B, Nitemberg G, *et al.* Prospective multicenter study of vascular-catheter-related complications and risk factors for positive central-catheter cultures in intensive care unit patients. J Clin Microbiol. 1990;28:2520-2525.

28. Pittet D. Intravenous catheter-related infections:current understanding. Paper presented at: Programs and Abstracts of the Thirty-second Interscience Conference on Antimicrobial Agents and Chemotherapy, 1992; Anaheim, CA.

29. Heard SO, Wagle M, Vijayakumar E, *et al.* Influence of triple-lumen central venous catheters coated with chlorhexidine and silver sulfadiazine on the incidence of catheter-related bacteremia. Arch Intern Med. 1998;158:81-87.

30. Merrer J, De Jonghe B, Golliot F, *et al.* Complications of femoral and subclavian venous catheterization in critically ill patients: a randomized controlled trial. JAMA. 2001;286:700-707.

31. Goetz AM, Wagener MM, Miller JM, Muder RR. Risk of infection due to central venous catheters: effect of site of placement and catheter type. Infect Control Hosp Epidemiol. 1998;19:842-845.

32. Tacconelli E, Tumbarello M, Pittiruti M, *et al.* Central venous catheter-related sepsis in a cohort of 366 hospitalised patients. Eur J Clin Microbiol Infect Dis. 1997;16:203-209.

33. Kealey GP, Chang P, Heinle J, Rosenquist MD, Lewis RW, 2nd. Prospective comparison of two management strategies of central venous catheters in burn patients. J Trauma. 1995;38:344-349.

34. Snyder RH, Archer FJ, Endy T, *et al.* Catheter infection. A comparison of two catheter maintenance techniques. Ann Surg. 1988;208:651-653.

35.   Cobb DK, High KP, Sawyer RG, *et al*. A controlled trial of scheduled replacement of central venous and pulmonary-artery catheters. N Engl J Med. 1992;327:1062-1068.

36.   Michel LA, Bradpiece HA, Randour P, Pouthier F. Safety of central venous catheter change over guidewire for suspected catheter-related sepsis. A prospective randomized trial. Int Surg. 1988;73:180-186.

37.   Eyer S, Brummitt C, Crossley K, Siegel R, Cerra F. Catheter-related sepsis: prospective, randomized study of three methods of long-term catheter maintenance. Crit Care Med. 1990;18:1073-1079.

38.   Oliver MJ, Callery SM, Thorpe KE, Schwab SJ, Churchill DN. Risk of bacteremia from temporary hemodialysis catheters by site of insertion and duration of use: a prospective study. Kidney Int. 2000;58:2543-2545.

39.   Pettigrew RA, Lang SDR, Haydock DA, Parry BR, Bremner DA, Hill GL. Catheter-related sepsis in patients on intravenous nutrition: a prospective study of quantitative catheter cultures and guidewire changes for suspected sepsis. Br J Surg. 1985;72:52-55.

40.   Savage AP, Picard M, Hopkins CC, Malt RA. Complications and survival of multilumen central venous catheters used for total parenteral nutrition. Br J Surg. 1993;80:1287-1290.

41.   Bach A, Stubbig K, Geiss HK. Infectious risk of replacing venous catheters by the guide-wire technique. Zentralbl Hyg Umweltmed. 1992;193:150-159.

42.   Senagore A, Waller JD, Bonnell BW, Bursch LR, Scholten DJ. Pulmonary artery catheterization: a prospective study of internal jugular and subclavian approaches. Crit Care Med. 1987;15:35-37.

43.   Powell C, Fabri PJ, Kudsk KA. Risk of Infection accompanying the use of single-lumen vs double-lumen subclavian catheters: a prospective randomised study. Journal of Parenteral and Enteral Nutrition. 1988;12:127-129.

44.   Bach A, Bohrer H, Geiss HK. Safety of a guidewire technique for replacement of pulmonary artery catheters. J Cardiothorac Vasc Anesth. 1992;6:711-714.

45.   Cook D, Randolph A, Kernerman P, *et al*. Central venous catheter replacement strategies: a systematic review of the literature. Crit Care Med. 1997;25:1417-1424.

46.   Hoffmann KK, Weber DJ, Samsa GP, Rutala WA. Transparent polyurethane film as an intravenous catheter dressing. A meta-analysis of the infection risks. JAMA. 1992;267:2072-2076.

47.   Widmer AF. Intravenous Catheter-related Infections. In: Wenzel RP, ed. Prevention and Control of Nosocomial Infections. 3rd edition ed. Baltimore, MD: Williams and Wilkins; 1997.

48.   Lederle FA, Parenti CM, Berskow LC, Ellingson KJ. The idle intravenous catheter. Ann Intern Med. 1992;116:737-738.

49.   Bregenzer T, Conen D, Sakmann P, Widmer AF. Is routine replacement of peripheral intravenous catheters necessary? Arch Intern Med. 1998;158:151-156.

50.   Uldall PR, Merchant N, Woods F, Yarworski U, Vas S. Changing subclavian haemodialysis cannulas to reduce infection. Lancet. 1981;1:1373.

Chapter 2

EPIDEMIOLOGY AND PATHOGENESIS OF CATHETER-RELATED BLOODSTREAM INFECTIONS

Antonio Sitges-Serra, F.R.C.S (Ed.)
Department of Surgery, Hospital Universitari del Mar, Barcelona, Spain

Introduction

As the only surgeon contributing to this major work on catheter-related bloodstream infections (CRBSI), this perhaps requires some explanation. In the mid- seventies, early during my residency training in general surgery, I was charged to take care of patients receiving parenteral nutrition at the Hospital de Bellvitge, my home institution, which pioneered this modality treatment in Spain (1). Our general surgery service was a reference one for the care of patients with complex postoperative abdominal complications such as fistulas and short bowel syndrome. Being a young trainee, I took this responsibility with some fear but with much enthusiasm because it gave me a unique opportunity both for challenging patient care and for first-line research. Of the many interesting aspects of parenteral nutrition delivery, subclavian catheter infections readily attracted my attention. They were common, they forced us to stop treatment and carried significant morbidity and even mortality. Initial attempts at controlling this complication by meticulous skin care and full barrier precautions at catheter insertion were unsuccessful and, at a given point, catheter sepsis due to coagulase-negative staphylococci attained almost an endemic proportion. This prompted us to start a series of investigations that led, in the early 1980s, to the recognition of the catheter hub as a relevant portal of entry for microorganisms contaminating central venous catheters (CVCs). The paper reporting this seminal observation could have appeared some years earlier but, unfortunately, it was rejected in three medical journals before being accepted in a journal specialised in artificial nutrition (2).

Our findings had a somehow cold reception in the expert entourage of the time probably because they came from an unknown unit and because they challenged established knowledge. In fact, intellectual challenge was welcomed by our group and represented a potent stimulus for us to hypothesize that the new paradigm of endoluminal catheter contamination would have a major impact on issues closely related to pathogenesis of CRBSI, namely, diagnosis, prevention, treatment and industrial design of future catheters (3). These thoughts have been largely confirmed by many research groups and have been the focus of our work spanning over twenty years (1976-1996) during which we made some contributions to the field and had the privilege to meet and discuss the issue with experts on both sides of the Atlantic.

DEFINITIONS

There is a growing consensus on the definitions to be used when dealing with infections due to intravascular devices. In the present chapter, we have adhered to a set of definitions recently proposed to improve communication between researchers and increase scientific accuracy of articles dealing with catheter infections (4).

1) The term intravascular catheter-related bloodstream infections (CRBSI) denotes bacteremia or fungemia in a patient who has an intravascular device, >1 positive blood culture from a peripheral vein, clinical manifestations of infection and confirming appropriate microbiological cultures. CRBSI is to be preferred to the term "catheter-related sepsis" since the concept "sepsis" does not imply bacteremia and is used to define the systemic inflammatory response syndrome associated with an infectious focus. "Catheter-related bacteremia" is not accurate since blood cultures may grow fungal species (fungemia).

2) Exit-site infection should be reserved to refer to clinical and/or microbiologically proven infection at the catheter exit site: periorificial cellulitis, purulence, tunnelitis, and pocket infections (for totally implantable devices). The term "infected catheter", usually employed to designate catheter segments yielding a high bacterial count in semiquantitative or quantitative cultures, should be abandoned.

3) Catheter colonization means that the cultured catheter segment (hub, tip, subcutaneous segment) grows a significant number bacteria according to the culture methods used.

To these we add the concept of catheter contamination to designate the process whereby microorganisms reach the catheter, adhere to its surface and proliferate. Thus, catheter contamination becomes the central focus of pathogenesis studies.

EPIDEMIOLOGY

Current care of severely ill patients requires multiple vascular accesses for invasive physiologic monitoring and intravenous delivery of fluids, drugs and parenteral nutrition. It is not uncommon for patients in the ICU to have at least three intravascular devices placed at the same time: one or two for hemodynamic monitoring and one or two for drugs and fluid therapy. Multilumen catheters may have reduced the overall number of lines used, but the increased number of hubs poses additional line management problems. These are the main reasons why intravascular devices are responsible for the ever-increasing prevalence of hospital-acquired gram-positive cocci bacteremia (5,6).

CRBSI represents the first cause of hospital acquired bacteremia with an approximate incidence of one episode per hundred hospital admissions (7). The incidence of CRBSI has been estimated to be in the range of 2 to 14 episodes for 1000 catheter-days (8,9). The rate of CRBSI for peripherally inserted lines and cannulas is lower than for central venous lines. In a recent survey at the Hospital del Mar (unpublished observations), 500 peripheral cannulas inserted in patients admitted to medical and surgical wards, with a mean indwelling time of 5.4 days, were prospectively investigated with skin, hub and tip semiquantitative cultures. Three (0.6%) were associated with bacteremia. However, colonization of the skin, hub and tip was found in roughly 18%, 9% and 18% of the cannulas, respectively. Midline catheters (n=215) with a longer dwelling time (7.2 days) were also investigated and showed a 2.5% CRBSI rate or 1.6 episodes /1000 catheter days.

The mean cost per CRBSI episode was calculated ten years ago to be roughly US$ 3,700 per episode with an average prolongation of hospital stay of one week (10). However, costs can be as high as US$ 6-7,000 for *Staphylococcus aureus* bacteremia which is often associated with distant septic metastasis and requires prolonged antibiotic therapy (5). A recent case-control study carried out in Spain (11), has confirmed these data. The average cost of an episode of CRBSI was found to be € 3,250 (year 1998). One-third of the patients were responsible for two-thirds of the increased total cost. The hospital stay for patients with CRBSI doubled that of the controls (26.5 vs 14.5 days).

CRBSI: A TIME-DEPENDENT PROCESS

Dwelling time is probably the most important factor influencing CRBSI rates. This is easily understood if we imagine two extreme scenarios:

1) A patient undergoes herniorraphy in an ambulatory surgical setting. Anesthetics and analgesics are given through a short peripheral cannula and the patient is discharged home the same day. The risk of CRBSI for this intravenous cannula is virtually zero.

2) A patient requires long-term parenteral nutrition for intestinal failure and receives home parenteral feeding through a tunnelled central line. He or she has an almost 100% possibility of suffering a catheter infection within the next two years.

Dwelling time varies widely between these two examples in clinical practice and represents a major conceptual obstacle to reach consensus on topics such as routes of catheter contamination, efficacy of preventive measures or sensitivity of diagnostics tests, to mention only a few areas in which controversy still persists. In addition, as discussed below, the relationship between dwelling time and CRSBI rates is not a linear one and, in fact, it appears to be bimodal. Thus, a thorough understanding of the relevance and subtleties of dwelling time has become essential to judge and to interpret data on catheter infections.

Overt clinical symptoms of CRBSI are preceded by catheter contamination, a complex process with a silent natural history that comprises several steps. First, microorganisms reach and contaminate the catheter segment(s); second, microorganisms adhere and proliferate on their surfaces; and third they seed the bloodstream once they have reached a significant number (colonized catheter). When the bloodstream is seeded, symptoms of bacteremia develop and, hopefully, the appropriate diagnosis is made. Thus, to properly understand the pathogenesis of CRBSI, it is essential to distinguish these different processes that develop over time in a sequential manner:

1) The routes of catheter contamination;
2) The interaction between microorganisms and catheter material;
3) The invasiveness of specific microorganisms.

In addition, some clinical/anatomical circumstances may facilitate this time-dependent process. Strictly speaking, these "risk factors", do not imply alternative contamination routes but, rather, exemplify how dwelling time, access site or dressing (to mention only a few examples) may predispose the catheter to get contaminated by a given route or by a specific bacteria or fungus.

ROUTES OF CATHETER CONTAMINATION

Microorganisms can reach the catheter's external surface (extraluminal contamination) either by migration from the skin entry site and progression along the subcutaneous tract or, more rarely, from a bacteremia stemming from a distant source (i.e., urinary tract infection, intraabdominal infection). Alternatively, microorganisms can reach the catheter's internal surface (endoluminal contamination) after colonizing the hub or, exceptionally, from a contaminated infusate.

Extraluminal (skin exit-site originated) contamination

This is the best known route of catheter contamination since it was described more than 40 years ago (12) and is the most relevant contamination route for catheters inserted for less than a week (13). Actually, the implication of skin microorganisms in CRBSI occurring early after insertion has been exhaustively documented both by conventional and molecular biology bacterial identification techniques (14). Furthermore, the mechanisms and speed of *S. aureus* migration along the subcutaneous tract have been investigated and clarified in an animal model (15).

Because of poor skin preparation, defective surgical technique, or inappropriate dressing of the fresh skin puncture, the catheter skin exit-site wound gets contaminated during catheter insertion or shortly afterwards by microorganisms of the skin flora. This contamination may progress to a subdermal infection, which spreads along the catheter track and reaches its intravascular segment and tip. Occasionally, a catheter exit-site infection may be the origin of a severe soft tissue infection (Figure 1) or a septic phlebitis. If the catheter has been tunneled to the anterior chest wall, insertion-site infection may give rise to a clinically evident soft-tissue infection, also called tunnel infection (4). The organisms most often involved are skin commensals such as coagulase-negative staphylococci and *S. aureus*, but in hospitalized patients the skin flora may also include other pathogens, such as *Enterococcus spp., Enterobacteriaceae,* or *Pseudomonas spp.*, all of which are also found in skin-originated CRBSI.

Experimental evidence suggests that, with time, the skin entry site and the subdermal tunnel become relatively resistant to the invasion by skin comensals (16) and this may explain why skin-related infections usually develop early after catheterization.

Extraluminal contamination is uncommon during intravenous feeding since TPN catheters are almost universally inserted using maximal aseptic barriers (17) Maximizing aseptic care at catheter insertion results in complete prevention or in very low rates of extraluminally originated CRBSI (18). This is most logical since this route of contamination has many similarities with that of a surgical clean wound infection and its prevention relies, as well, in adopting maximal aseptic barriers at the time of catheter insertion which should be considered as a minor surgical procedure.

Extraluminal infection can also occur in patients with a bacteremia from a distant source. Microorganisms in the circulation may adhere to the intravascular catheter segment and seed the bloodstream from this secondary "metastatic" septic focus. This possibility should be borne in mind particularly in critically ill patients with persistent bacteremia after an apparently successful treatment of an obvious septic focus.

Endoluminal (hub originated) contamination

For decades experts held the opinion that catheters were contaminated almost exclusively by bacteria or fungi present at the skin's catheter exit-site. Two other routes of contamination were exceptionally considered: intravascular device contamination from bacteremia arising from a distant focus (hematogenous seeding) or from a contaminated infusate. The "skin paradigm" was born in the 60's, when investigators became aware of the severity of CRBSI due to *S. aureus* (12). It became firmly established after the description of the semiquantitative catheter tip culture method based on culturing the external surface of the catheter tip (19). The widespread belief in the skin paradigm, however, prevented many investigators from appropriately appraising some clinical observations that did not match with this contamination route:

1) Many patients with CRBSI have no clinical infection at the catheter skin entry site;
2) Mutbreaks of CRBSI were reported in relationship to loosening of the catheter-infusion set junction (20) This was initially attributed to moistening of the dressing and subsequent extraluminal contamination.
3) In a substantial proportion of CRBSI, the microorganism recovered from blood and catheter tip cultures did not match with that isolated from the skin entry site (21).
4) Locking the line with heparinized saline would often result in clearance of fever and chills.

These facts were appropriately interpreted once the relevance of endoluminal contamination was recognized in the mid-eighties in relationship to CRBSI originated from parenteral nutrition catheters (2,17). At that time, outbreaks of CRBSI due to coagulase negative staphylococci had been reported (22) usually during the second to fourth weeks after catheter placement. Symptoms frequently vanished after locking the catheter and stopping parenteral nutrition. In our institution, this outbreak could not be controlled by inserting the subclavian catheters in the operating room, by tunnelizing the subclavian lines, nor by improving skin antisepsis.

We then started a series of studies that included separate quantitative cultures of the inner surface of the catheter hub and segments (Figure 2). In a first study (17) episodes of CRBSI were investigated with a multiple culture protocol including sampling of peripheral blood, hub, inner and outer catheter tip surfaces, skin entry site, parenteral nutrition mixture and distant infected sites (2) Results clearly indicated that in the majority of cases the microorganisms present at the inner hub surface were the same (species and antibiotype) as those recovered from the catheter tip and blood. Further studies in a larger series of parenteral nutrition subclavian catheters, inserted for a mean of three weeks, confirmed the relevance of the endoluminal contamination route which accounted for 70% of all CRBSI (17) The skin, the all-in-one nutrient mixtures and hematogenous seeding of the intravascular segment accounted for the remaining 30%. Bacteriological findings also demonstrated that when the hub was involved, the same microorganisms were recovered from the inner catheter surface at the middle third and at its tip, indicating that the whole catheter lumen was seeded with bacteria stemming from the proximal hub.

In 1992, the first North American paper recognizing the relevance of endoluminal contamination was published. Salzman *et al.* (23) carried out hub cultures from long-term central lines in neonates and found microorganisms matching those recovered from blood cultures and the catheter tip. Recognition of hub colonization was delayed in the USA mainly for two reasons: reluctance to implement hub cultures and scheduled line replacements in the intensive care unit (ICU), a widespread empirical practice with little scientific basis (24) but that may have reduced the chances of hub colonization by reducing the mean catheter dwelling time. Currently, the role of hubs as portals of entry for microoganisms is widely accepted although there is still an ongoing controversy on the relative importance of this route of contamination as opposed to the skin. Evidence derived from microscopic examination of catheter surfaces (25) and data coming from series of CRBSI in which hubs have been cultured (26) indicate that the longer the catheters remain in place the more likely they are to become contaminated endoluminally (Table 1).

Table 1. Relative importance of different routes of contamination of intravascular catheters.

Author	Catheters N	Dwell Time (d)	CRBSI	Hub	Skin	Mixed	Other[a]
Cercendado (27)	139	8.6	53[b]	12	30	8	3
Llop (8)	3632	13	179	50	28	35	66
Fan (28)	156	15	11	1	4	2	4
Cicco (29)	109	18.2	6	3	3	-	-
Liñares (17)	22	20	20	14	2	-	4
Segura (30)	400	23	24	9	5	2	8
Salzman (23)	113	23.9	28	21	7	-	-
Weightman (31)	42	114	11	8	-	-	3[c]

a)   Hematogenous seeding, infusate contamination, unknown.
b)   Positive catheter tips
c)   Incomplete cultures.

TYPE OF MICROORGANISMS AND INTERACTION WITH CATHETER MATERIAL

There are particular local or systemic circumstances facilitating catheter colonization and/or bloodstream seeding by a specific microorganism. In a colonization study of 3,632 parenteral nutrition central lines (mean dwelling time of 13 days), Llop *et al..* (8) were able to identify factors favoring catheter colonization by each of the microorganisms most commonly involved in CRBSI: coagulase-negative staphylococci, *S. aureus*, gram negative rods and fungi.

Coagulase-negative staphylococci were involved in 487 (60%) of the 823 colonized catheters. The rate of colonization/bacteremia for these bacteria, however, was the lowest: only 82 out of these 487 colonized catheters resulted in CRBSI. The corresponding figure for gram negative rods was 51/146; for fungi 13/22 and for S. aureus 25/102. Other microorganisms such as streptococci, *Corynebacterium spp.*, *Bacillus spp*, and enterococci, showed a similarly low degree of bloodstream invasiveness (13/91). Thus, not all microorganisms colonizing catheter tips exhibit the same potential for symptomatic blood seeding.

Fungal catheter colonization had the strongest association with dwelling time, possibly because extensive skin, pharyngeal or gastrointestinal colonization by Candida spp. in the non-immunocompromised host tends to appear relatively late in the course of the disease, usually as a consequence of repeated abdominal surgery and prolonged antibiotic treatment.

Catheter colonization by gram negative rods and fungi, but not by coagulase-negative staphylococci, was strongly associated (OR >7) with the presence of a distant septic focus. The reasons for these appear to be multiple: skin colonization by microorganisms present in the distant focus, hematogenous seeding of the catheter, selection of flora by the associated antibiotic treatment and cross-contamination during the manipulation of the catheter hub.

The interaction of microorganisms with catheter material has been the subject of much interest. Material rugosity, chemical composition and biofilm formation are some of the most relevant research areas in this field. In experimental studies (16) silicone catheters have been shown to elicit more inflammatory changes in the soft tissues surrounding the catheter and to facilitate *S. aureus* infections. There is no evidence, however, that catheter material has a measurable impact on CRBSI rates in humans.

The case of the coagulase-negative staphylococci. *S. epidermidis* and other coagulase-negative staphylococci such as *S. hominis* or *S. haemolyticus*, are responsible for about two-thirds of CRBSI. They represent the main skin commensals (not only in patients but also on the hands of healthcare workers!) and have a particular facility to adhere and replicate on plastic surfaces. The mechanisms of coagulase-negative staphylococci contamination are starting to be unveiled. Atela *et al.* (32) investigated the dynamics of catheter segment contamination by these bacteria using strain delineation and reached important conclusions. They were able to show that:

1) Contamination of external catheter segments and skin entry site is often transient either because of the biological characteristics of the microorganisms or the effects of line manipulation and dressing changes;
2) The same microorganisms could be found in superficial cultures of hub or skin in less than 30% of catheters with positive tips for coagulase-negative staphylococci;
3) Around of 80% of catheter contaminations by the coagulase-negative staphylococci were polyclonal;
4) Two-thirds of instances of hub contamination took place during the first 10 days after insertion.

Thus, it appears that different strains of coagulase-negative staphylococci can be found in superficial cultures soon after catheter insertion, that permanent colonization is not the rule and that polyclonality is so common that the usefulness of strain delineation for the diagnosis of CRBSI can be challenged. Polyclonality has also been described in endocarditis due to *S. epidermidis* (33) and adds further difficulties to the understanding of these infections.

The mechanisms of bacterial adherence

Bacterial adherence on catheter surfaces is a complex phenomenon resulting from an interplay of at least three factors: the catheter material (rugosity and polarity), the host response (biofilm formation) and bacterial adhesion factors. Much progress has recently been made on the mechanisms whereby bacteria adhere to foreign body surfaces, a process designed to anchor them in a nutritionally advantageous environment in which, in addition, they become protected from host defenses and antibiotics.

Early *in vivo* work with the scanning and transmission electron microscope (Figure 3) demonstrated that microorganisms become buried within the pits and creaks of the very irregular intravenous catheter surfaces, often between epithelial cells desquamated from the skin at the time of catheter insertion (34) This process is much facilitated by the biofilm generated by the host and also by the production of adhesins and mucoid substances by the bacteria and fungi themselves.

Biofilm formation is largely a response of the host against a foreign body and appears to be facilitated by rough surfaces and hydrophobic materials. It essentially consists of the deposition of proteins on the surface of the device in which epithelial cells, inflammatory cells and the microorganisms themselves become trapped. Host proteins involved are predominantly fibrin and fibronectin. To these, multiple bacterial products, broadly referred to as extracellular polymeric substances, are added on. Furthermore, the infusate itself, particularly if it contains TPN mixtures, can leave residue on the surface of the catheter. The resulting structure of the biofilm is not a mere homogeneous monolayer of slime but is a heterogeneous multilayer habitat, both in space and over time, with "water channels" that allow transport of essential nutrients and oxygen to the cells and microorganisms growing within the biofilm (35) The progression leading to a mature biofilm requires changes in bacterial gene expression which seem to be induced by environmental stimuli.

RISK FACTORS FOR CATHETER INFECTIONS

Some clinical variables influence the rates of CRBSI by favoring catheter contamination by either the extraluminal or endoluminal routes. These have been recently reviewed in depth by Safdar *et al.*. (36), thus only a focused overview is presented here.

Indwelling time. Indwelling time has been repeatedly shown to be one of the most important risk factors for CRBSI (36-38). The relationship between dwelling time and CRBSI can be described as bimodal, at least in the hospitalized patient. There seems to be an incidence peak early after catheterization (<10 days) in relation with extraluminal contamination and an exit-site infection and, and a second one occurring after the second catheterization week, which most usually represents contamination by the endoluminal route. The higher rates of CRBSI associated with TPN or hemodialysis catheters are probably due to a long indwelling time rather than to specific local or systemic factors.

Access site. Several studies have reported increased CRBSI rates for the jugular and femoral approaches compared to the subclavian or peripheral insertion accesses (36-38). In the study of Llop *et al.*. (8), the internal jugular access showed the highest colonization and CRBSI risks for the coagulase-negative staphylococci, *S. aureus* and gram negative rods. Increased colonization of internal jugular catheters may be explained by the rugosity of the skin area, problems with catheter and dressing fixation and the growing beard, in male patients. Tunneling internal jugular catheters to the subclavian area has shown significant benefit in reducing CRBSI rates (39) and this is probably not due to the effect of simply prolonging the subcutaneous tract but rather to converting a jugular skin entry-site into a subclavian site. Thus, from extensive epidemiological studies it can be concluded that the risk of catheter colonization according to access sites in an ascendent order would be subclavian = basilic < femoral < internal jugular.

Number of hubs

Multiple lumen catheters have been associated with an increased infection rates in some studies (40-42) but not in others (43-44) Since any increase of the CRBI rates for multiple lumen catheters should be ascribed to the increased likelihood of endoluminal contamination, studies such as that of Gil *et al.* (43). with a mean indwelling time of less than a week may not appropriately reflect the higher potential for CRBSI of these devices. With increased dwelling time, however, the number of hubs can make a difference. The study of Llop *et al.*. (8) sheds new light on this controversy. In their study, an increasing number of lumens tended (P=0.05) to facilitate colonization by the coaguase-negative staphylococci, a group of microorganisms particularly involved in endoluminal contamination. For other microorganisms the number of lumens had no impact on colonization rates either because the sample was small (fungi; P=0.1) or because they are commonly involved in extraluminal contamination (gram negative rods or *S. aureus*).

Parenteral nutrition and infusate-related CRBSI

TPN has not been identified as an independent risk factor for CRBSI in multivariate analysis (37). Increased CRBSI rates attributable to lipid administration was suggested in one epidemiological study in which *S. epidermidis* bacteremia was detected more often in neonates receiving lipids (45). The most probable culprit of these high CRBSI rates, however, was the lipid administration system favoring endoluminal contamination and not fat per se (46). Thus, the higher risk of CRBI occasionally found in patients on

TPN can be attributed to confounding factors such as the catheterization time, which is superior for TPN catheters than for other types of lines, and frequent manipulation of hubs.

Aseptic mixing of different TPN components is essential to avoid infusate-related CRBSI that can be fatal. Many environmental (mesophilic) microorganisms such as *Candida spp.*, coagulase negative staphylococci, *Bacillus spp.*, *E. cloacae*, or *E. coli* can grow luxuriously in TPN bags at room temperature. Bacterial growth in TPN solutions, however, is very sensitive to temperature (47). At 4°C all bacterial growth ceases and, for this reason, storage of up to a week of TPN mixtures is allowed at this temperature (48). Some psycrophilic bacteria, however, such as *P. cepacia*, *P. fluorescens* or *Stenotrophomonas maltophilia* grow at rather low temperatures (8°C) and their isolation from blood in patients receiving TPN should alert the physician to look for potential infusate contamination.

Understaffing and education

The role of proper training of personnel handling central lines cannot be overemphasized. In the Fridkin *et al.*. case-control study (49) the prevalence of CRBSI almost doubled during a time period in which a parallel decrease in the number of nurses took place. Understaffing was identified by these authors as a major circumstance predisposing to CRBSI. Interestingly, this outbreak could not be controlled by introducing antibiotic-coated catheters, thus emphasizing, once more, that present day technology cannot be considered as a substitute for the time-honored principles of catheter insertion and care. For this reason, we cannot agree with the current CDC recommendation (50) that antibiotic-coated catheters should be used "when after implementing a comprehensive strategy to reduce rates of CRBSI, the CRBSI rates remain above the goal set by the individual institution". This is a misleading sentence that suggests an unrealistic solution.

Finally, education and proper training will have an increasing importance in the times to come. Frequent rotations, the disappearance of intravenous teams, and leaving responsibility for vascular access to young trainees, are common circumstances facilitating suboptimal catheter insertion and hub handling. The issue has been extensively demonstrated in two recent outstanding articles that stress the importance of proper training and continuous education at the institutional level as a protection variable against CRBSI (51-52).

ACKNOWLEDGEMENTS

For twenty years I have shared my work with so many colleagues that there is no way I can do justice to all of them in a few lines. Studies carried out at the Hospital de Bellvitge could not have been performed without the help of J. Liñares, a most learned and meticulous clinical microbiologist, with whom I have kept a lifetime friendship. The development of an animal model of hub-related CRBSI and the design and testing of a new antiseptic hub was largely the result of the collaboration with M. Segura and C. Alía at the Hospital del Mar. Since the early 1990s, many other Spanish groups have made important contributions to the field of CRBSI and I have particularly enjoyed exchanging my views with G. Prats, F. Álvarez, J. Rello, R. Pallarés, X. Garau, E. Bouza, C. León and J. Capdevila. I am also indebted to many authors of this book for their insightful advice and positive criticism of our work, and, in particular, to the Editors, N. P. O'Grady and D. Pittet, for asking me to contribute. The Beatles (53) inspired in me the title of a newsbreaking letter to *The Lancet* (54).

REFERENCES

1.   Sitges-Creus A. Manual de Alimentación Parenteral. Ed. Toray-Masson, 1978. Bacelona, Spain.
2.   Sitges-Serra A, Puig P, Liñares J, Pérez JL, Farreró N, Jaurrieta E, Garau J. Hub colonization as the initial step in an outbreak of catheter-related sepsis due to coagulase negative staphylococci during parenteral nutrition. JPEN 1984; 8:668-72.
3.   Sitges-Serra A, Liñares J, Garau J. Catheter sepsis: The clue is the hub. Surgery 1985; 97:355-7.
4.   Mermel LA, Farr BM, Sherertz RJ, Raad II, O'Grady N, Harris JS, Craven DE. Guidelines for the management of intravascular catheter-related infections. Clin Infect Dis 2001; 32:1249-72.
5.   Thylefors JD, Harbarth S, Pittet D. Increasing bacteremia due to coagulase-negative staphylococci: fiction or reality? Infect Control Hosp Epidemiol 1998; 19:581-9.
6.   Vallés J, León C, Álvarez-Lerma F. Nosocomial bacteremia in critically ill patients: a multicenter study evaluating epidemiology and prognosis. Clin Infect Dis 1997; 24:387-95.
7.   Widmer A. I.V.-related infections. In Wenzel RP "Prevention and control of nosocomial infections". Williams and Wilkins, Baltimore, 1993. pp. 556-79.

8.   Llop J, Badía MB, Comas D, Tubau M, Jodar R. Colonization and bacteremia risk factors in parenteral nutrition catheterization. Clin Nutr 2001; 20:527-34.
9.   Eggimann P, Pittet D. Overview of catheter-related infections with special emphasis on prevention based on educational programs. Clin Microbiol Infect 2002; 8:295-309.
10.   Arnow PM, Quimosing EM, Beach M. Consequences of intravascular catheter sepsis. Clin Inf Dis 1993; 16:778-84.
11.   Morís de la Tassa J, Fernández-Muñoz P, Antuña A, Gutiérrez del Río MC, de la Fuente B, Cartón J. Estudio de los costes asociados a la bacteriemia relacionada con el catéter. Rev Clin Esp 1998; 198:641-6.
12.   Crane Ch. Venous interruption for septic thrombophlebitis. N Engl J Med 1960; 262:947-51.
13.   Maki DG. Pathogenesis, prevention and management of infections due to intravascular devices used for infusion therapy. In: "Infections associated with indwelling medical devices". Bisno AL, Waldvogel FA (eds.). American Society for Microbiology, Washington DC, 1989, pp.171-7.
14.   Maki DG. Infections due to infusion therapy. In: "Hospital Infections", Bennet JV and Brachman PS (eds.). Little, Brown and Company, Boston, 1992, pp. 849-98.
15.   Cooper GL, Schiller AL, Hopkins CC. Possible role of capillary action in pathogenesis of experimental catheter-associated dermal tunnel infections. J Clin Microbiol 1988; 26:8-12.
16.   Sherertz RJ, Carruth WA, Marosok RD, Espeland MA, Johnsson RA, Solomon DD. Contribution of vascular catheter material to the pathogenesis of infection: the enhanced risk of silicone in vivo. J Biomed Mat Res 1995; 29:635-45.
17.   Liñares J, Sitges-Serra A, Garau J, Pérez JL, Martín R. Pathogenesis of catheter sepsis: a prospective study with quantitative and semiquantitative cultures of catheter hub and segments. J Clin Microbiol 1985; 21:357-60.
18.   Raad II, Hohn DC, Gilbreath J, Suleiman N, Hill LA, Bruso PA, Marts K, Mansfiel P, Bodey GP. Prevention of central venous catheter-related infections by using maximal sterile barrier precautions during insertion. Inf Control Hosp Epidemiol 1994; 15:231-8.
19.   Maki DG, Weise CE, Sarafin HW. A semiquantitative culture method for identifying intravenous catheter related infection. N Engl J Med 1977; 296:1305-9.
20.   Deitel M, Krajden S, Saldanha CF, Gregory WD, Fuksa M, Cantwell E. An outbreak of Staphylococcus epidermidis septicemia. JPEN 1983; 7:569-72.
21.   Bjornson HS, Colley R, Bower RH, Duty VP, Schwartz-Fulton JT, Fischer JE. Association between microorganism growth at the catheter insertion site and colonization of the catheter in patients receiving total parenteral nutrition. Surgery 1982; 92:720-7.
22.   Sitges-Serra A, Puig P, Jaurrieta E, Garau J, Alastrue A, Sitges-Creus A. Catheter sepsis due to Staphylococcus epidermidis during parenteral nutrition. Surg Gynecol Obstet 1980; 151:481-3.
23.   Salzman MB, Isenberg HD, Shapiro JF, Lipsitz PJ, Rubin LG. A prospective study of the catheter-hub as the portal of entry for microorganisms causing catheter-related sepsis in neonates. J Infect Dis 1993; 167:487-90.

24. Cobb DK, High KP, Sawyer RG A controlled trial of scheduled replacement of central venous and pulmonary-artery catheters. N Engl J Med 1992; 327:1062-8.

25. Raad I, Costerton JW, Sabharwall U, Sacilowski M, Anaissie E, Bodey GP. Ultrastructural analysis of indwelling vascular catheters: A quantitative relationship between luminal colonization and duration of placement. J Inf Dis 1993; 168:400-7.

26. Sitges-Serra A, Hernández R, Maestro S, Pi-Suñer T, Garcés JM, Segura M. Prevention of catheter sepsis: the hub. Nutrition 1997; 13(suppl.):30-5S.

27. Cercenado E, Ena J, Rodríguez-Créixems M, Romero I, Bouza E. A conservative procedure for the diagnosis of catheter-related infections. Arch Int Med 1990; 150:1417-20.

28. Fan ST, Teoh-Chan CH, Lau KF, Chu KW, Kwan AKB, Wong KK. Predictive value of surveillance skin and hub cultures in central venous catheter sepsis. J Hosp Infection 1988; 12:191-8.

29. Cicco M, Panarello G, Chiaradia V, Fracasso A, Veronesi A, Testa V, Santini G, Tesio F. Source and route of microbial colonisation of parenteral nutrition catheters. Lancet 1989; 2:1258-60.

30. Segura M, Lladó L, Guirao X, Piracés M, Herms R, Alia C, Sitges-Serra A. A prospective study of a new protocol for in situ diagnosis of central venous catheter related bacteraemia. Clin Nutr 1993; 12:103-7.

31. Weightman NC, Simpson EM, Spelkler DCE, Mott MG, Oakhill A. Bacteraemia related to indwelling central venous catheters: Prevention, diagnosis and treatment. Eur J Clin Microbiol 1988; 7:125-9.

32. Atela I, Coll P, Rello J, Quintana E, Barrio J, March F, Sánchez F, Barraquer P, Ballús J, Cotura A, Prats G. Serial surveillance cultures of skin and catheter hub specimens from critically ill patients with central venous catheters: Molecular epidemiology of infection and implications for clinical management and research. J Clin Microbiol 1997; 35:1784-90.

33. Van Wijngaerden E, Peetermans WE, Van Lierde S, Van Eldere J. Polyclonal Staphylococcus endocarditis. Clin Infect Dis 1997; 25:69-71.

34. Marrie ThJ, Costerton JW. Scanning and transmission electron microscopy of in situ bacterial colonization of intravenous and intrarterial catheters. J Clin Microbiol 1984; 19:687-93.

35. Donlan RM. Biofilm formation: A clinically relevant microbiological process. Clin Infect Dis 2001;33:1387-92.

36. Safdar N, Kluger DM, Maki DG. A review of risk factors for catheter-related bloodstream infection caused by percutaneously inserted, noncuffed central venous catheters: implications for preventive strategies. Medicine (Baltimore) 2002; 81:466-79.

37. Moro ML, Franco E, Cozzi A and The central venous catheter-related infections study group. Risk factors for central venous catheter-related infections in surgical and intensive care patients. Inf Control Hosp Epidemiol 1994; 15:253-64.

38. Richet H, Hubert B, Nitemberg G, Andremont A, Buu-Hoi A, Ourbak P, Galicier C, Veron M, Boisivon A, Bouvier AM, Ricome JC, Wolff MA, Pean Y, Berardi L, Bourdain JL, Hautefort B, Laaban JP, Tillant D. Prospective multicenter study of

vascular-catheter-related complications and risk factors for positive central-catheter cultures in intensive care unit patients. J Clin Microbiol 1990; 28:2520-5.

39. Timsit JF, Sebille V, Farkas JC, Misste B, Martin JB, Chevret S, Carlet J. Effect of subcutaneous tunneling on internal jugular catheter-related sepsis in critically ill patients. JAMA 1996; 276:1416-20.

40. Clark-Christoff N, Watters VA, Sparks W, Snyder P, Grant JP. Use of triple-lumen catheters for administration of total parenteral nutrition. JPEN 1992; 16:403-10.

41. McCarthy MC, Shives JK, Robison RJ, Broadie TA. Prospective evaluation of single and triple lumen catheters in total parenteral nutrition. JPEN J Parenter Enteral Nutr 1987; 11:259-62.

42. Pemberton LB, Lyman B, Lander V, Covinsky J. Sepsis from triple- vs single-lumen catheters during total parenteral nutrition in surgical or critically ill patients. Arch Surg 198; 121:591-4.

43. Gil RT, Kruse JA, Thill-Baharozian MC, Carlson RW. Triple- vs single-lumen central venous catheters. A prospective study in a critically ill population. Arch Intern Med 1989; 149:1139-43.

44. Farkas JC, Liu N, Bleriot JP, Chevret S, Goldsterin FW, Carlet J. Single- versus triple-lumen central catheter-related sepsis: a prospective randomized study in a critically ill population. Am J Med 1992; 93:277-82.

45. Freeman J, Goldmann DA, Smith NE, Sidebottom DG, Epstein MF, Platt R. Association of intravenous lipid emulsion and coagulase-negative staphylococcal bacteremia in neonatal intensive care units. N Engl J Med 1990; 323:301-8.

46. Pérez JL, Liñares J, Pallarés R, Sitges-Serra A, Martín R. Lipid emulsions and bacteremia in the neonatal intensive care unit (letter). N Engl J Med 1991; 324:267.

47. Jeppson R, Johansson M, Tengborn J. Bacterial growth properties in refrigerated all-in-one TPN mixtures. Clin Nutr 1987; 6:25-9.

48. Weil DC, Arnow PM. Safety of refrigerated storage of admixed parenteral fluids. J Clin Microbiol 1988; 26:1787-90.

49. Fridkin SK, Pear SM, Williamson TH, Galgiani JN, Jarvis WR. The role of understaffing in central venous catheter-associated bloodstream infections. Infect Control Hosp Epidemiol 1996; 17:150-8.

50. CDC and Prevention Guidelines for the Prevention of Catheter-related Infections MMWR 2002; Vol. 51(No. RR-10).

51. Sherertz RJ, Ely EW, Westbrook DM, Gledhill KS, Streed SA, Kiger B, Flynn L, Hayes S, Strong S, Cruz J, Bowton DL, Hulgan T, Haponik EF. Education of physicians-in-training can decrease the risk for vascular catheter infection. Ann Intern Med 2000; 132:641-8.

52. Eggimann P, Harbarth S, Constantin MN, Touveneau S, Chevrolet JC, Pittet D. Impact of a prevention strategy targeted at vascular-access care on incidence of infections acquired in intensive care. Lancet 2000; 355:1864-8.

53. Eleanor Rigby. Lennon J, McCartney P. In: "Revolver", EMI Parlophone,1966.

54. Sitges-Serra A, Liñares J. Bacteria in total parenteral nutrition catheters: Where do they (all) come from?. Lancet 1983; 1:531.

Chapter 3

DIAGNOSIS

Stephen O. Heard, M.D., F.C.C.M.
Department of Anesthesiology, University of Massachussetts Medical Center, Worcester, Massachusetts

Introduction

The diagnosis of catheter-related infection can be a challenge. Frequently, the clinician is presented with a febrile patient without another apparent source of infection. The question then arises: Should the catheter be removed? Studies and recent guidelines from the Healthcare Infection Control Practices Advisory Committee (HICPAC) and the Centers for Disease Control (CDC) suggest that fever alone should not be an indication for catheter removal (1). This chapter reviews the available methods to document catheter infection either by catheter removal or by using techniques that attempt to determine infection with the catheter left in place.

Implicit in the diagnosis of catheter-related infection is a standard, widely accepted definition. Unfortunately, there is no "gold standard" definition and confusion regarding clinical definitions and research definitions exist (2). Table 1 is one proposed set of definitions (1).

An additional problem associated with the diagnosis of catheter-related infection and catheter-related bloodstream infection (CRBSI) is the reliance on antibiograms (antibiotic susceptibility profiles) to determine if bacteria from the

Table 1. Definitions Of Catheter-Related Infection And Catheter-Related Bloodstream Infections (1).

Table I

Examples of Clinical Definitions For Catheter-Related Infections	Surveillance Definitions for Primary Bloodstream Infections (BSI), National Nosocomial Infections Surveillance System
Localized Catheter Colonization Significant growth of a microorganism (e.g. >15 colony forming units (CFU)) from the catheter tip, subcutaneous segment of the catheter, or catheter hub	**Laboratory-Confirmed BSI** Should meet at least one of the following criteria: *Criterion 1:* Patient has a recognized pathogen cultured from one or more blood cultures, and the pathogen cultured from the blood is not related to an infection at another site.
Exit Site Infection Erythema or induration within 2 cm of the catheter exit site, in the absence of concomitant bloodstream infection (BSI) and without concomitant purulence	*Criterion 2:* Patient has at least one of the following signs or symptoms: (>100.4°F [>38°C]), chills, or hypotension, and at least one of the following:
Clinical Exit Site Infection (Or Tunnel Infection) Tenderness, erythema, or site induration >2 cm from the catheter site along the subcutaneous tract of a tunneled (e.g. Hickman or Broviac) catheter, in the absence of concomitant BSI	1. Common skin contaminant (e.g., diphtheroids, *Bacillus* spp., *Propionibacterium* spp., coagulase-negative staphylococci, or micrococci) cultured from two or more blood cultures drawn on separate occasions.
Pocket Infection Purulent fluid in the subcutaneous pocket of a totally implanted intravascular catheter that might or might not be associated with spontaneous rupture and drainage or necrosis of the overlaying skin, in the absence of concomitant BSI	2. Common skin contaminant (e.g., diphtheroids, *Bacillus* spp., *Propionibacterium* spp., coagulase-negative staphylococci, or micrococci) cultured from at least one blood culture from a patient with an intravenous catheter, and the physician institutes appropriate antimicrobial therapy.
Infusate-Related BSI Concordant growth of the same organism from the infusate and blood cultures (preferably percutaneously drawn) with no other identifiable source of infection	3. Positive antigen test on blood (e.g., *Hemophilus influenzae, Streptococcus pneumoniae, Neisseria meningitides,* or group B streptococcus).
Catheter-Related BSI Bacteremia/fungemia in a patient with an intravascular catheter with at least one positive blood culture obtained from a peripheral vein, clinical manifestations of infections (i.e., fever, chills, and/or hypotension), and no apparent source for the BSI except the catheter. One of the following should be present: a positive semiquantitative (>15 CFU/catheter segment) or quantitative (>10³ CFU/catheter segment) catheter culture whereby the same organism (species and antibiogram) is isolated from the catheter segment and peripheral blood; simultaneous quantitative blood cultures with a ≥5:1 ratio between catheter and peripheral blood cultures; time difference of > 2 hours between a catheter blood culture and peripheral blood culture.	and signs and symptoms with positive laboratory results are not related to an infection at another site. *Criterion 3:* Patient aged <1 year has at least one of the following signs or symptoms: fever (>100.4°F [>38°C]), hypothermia (<98.6° F [<37° C]), apnea, or bradycardia, and at least one of the following: 1. Common skin contaminant (e.g., diphtheroids, *Bacillus* spp., *Propionibacterium* spp., coagulase-negative staphylococci, or micrococci) cultured from two or more blood cultures drawn on separate occasions. 2. Common skin contaminant (e.g., diphtheroids *Bacillus* spp., *Propionibacterium* spp., coagulase-negative staphylococci, or micrococci) cultured from at least one blood culture from a patient with an intravenous catheter, and the physician institutes appropriate antimicrobial therapy. 3. Positive antigen test on blood (e.g., *Hemophilus influenzae, Streptococcus pneumoniae, Neisseria meningitides,* or group B streptococcus). and signs and symptoms with positive laboratory results are not related to an infection at another site.
Clinical Sepsis Should meet at least one of the following criteria: *Criterion 1:* Patient has at least one of the following clinical signs with no other recognized cause: fever (>100.4°F [>38°C]), hypotension (systolic pressure <90 mm Hg), or oliguria (<20 mL/hr), and blood culture not done or no organisms or antigen detected in blood and no apparent infection at another site, and physician institutes treatment for sepsis. *Criterion 2:* Patient aged <1 year has at least one of the following clinical signs or symptoms with no other recognized cause: fever (>100.4°F [>38°C]), apnea, or bradycardia, and blood culture not done or no organisms or antigen detected in blood and no apparent infection at another site, and physician institutes treatment for sepsis.	**Arterial or Venous Infection** Included are arteriovenous graft, shunt, fistula, or intravenous cannulation. Should meet at least one of the following criteria: *Criterion 1:* Patient has organisms cultured from arteries or veins removed during a surgical operation and blood culture not done or no organisms cultured from blood. *Criterion 2* Patient has evidence of arterial or venous infection seen during a surgical operation or histopathologic examination. *Criterion 3:* Patient has at least one of the following signs or symptoms with no other recognized cause: fever (>100.4°F [>38°C]), pain, erythema, or heat at involved vascular site and >15 CFUs cultured from an intravascular cannula tip using a semiquantitative culture method and blood culture not done or no organisms cultured from blood.
Catheter-Associated BSI Defined by the following: • Vascular access device that terminates at or close to the heart or one of the great vessels. An umbilical artery or vein catheter is considered a central catheter. • BSI is considered to be associated with a central catheter if the catheter was in use during the 48-hour period before development of the BSI. If the time interval between onset of infection and device use is >48 hours, there should be compelling evidence that the infection is related to the central catheter.	*Criterion 4:* Patient has purulent drainage at the involved vascular site and blood culture not done or no organisms cultured from blood. *Criterion 5:* Patient aged <1 year has at least one of the following signs or symptoms with no other recognized cause: fever (>100.4°F [>38°C]), hypothermia (<98.6° F [<37° C]), apnea, bradycardia lethargy, or pain, erythema or heat at involved vascular site and >15 colonies cultured from intravascular cannula tip using semiquantitative method and blood culture not done or no organisms cultured from blood.

blood and catheter are the identical. A number of studies using DNA typing of bacteria (particularly coagulase negative staphylococci) have indicated that using the antibiogram to validate a clonal etiology of the strains is not a valid nethodology (3). Hence, many times, the incidence of CRBSI will be overestimated if the susceptibility profile alone is used.

DIAGNOSIS OF CATHETER INFECTION WITH REMOVAL OF THE CATHETER

Semiquantitative Cultures

Over 25 years ago, catheters suspected of causing infection were removed and cultured qualitatively in a broth solution. Catheters were classified as being either positive or negative for growth. Such a practice produced poor sensitivity and specificity for infection. In 1977, a semiquantitative technique was described for culturing catheter tips (4). Catheters were aseptically removed from the patient and the tip (5 cm) of the catheter was cut and rolled 4 times across a blood agar plate. After 24-48 hours of incubation, colony-forming units (CFU) were counted. There was an association between a catheter yielding greater than or equal to 15 CFUs and inflammation at the insertion site. In addition, confluent growth on the blood agar plate occurred only in those patients with CRBSI. Unfortunately, the positive predictive value of the 15 CFU cutoff for CRBSI was only 16%. Nonetheless, because of its simplicity and reliability, the roll plate technique is the standard for diagnosing catheter-related infection and a catheter with a CFU count of greater than or equal to 15 is considered to be significantly colonized. In addition, a recent meta-analysis evaluating the utility of catheter-tip colonization (both semiquantitative and quantitative [vide infra]) as a proxy end point for CRBSI revealed a fairly good correlation between the two (Figure 1, r=0.69) (5).

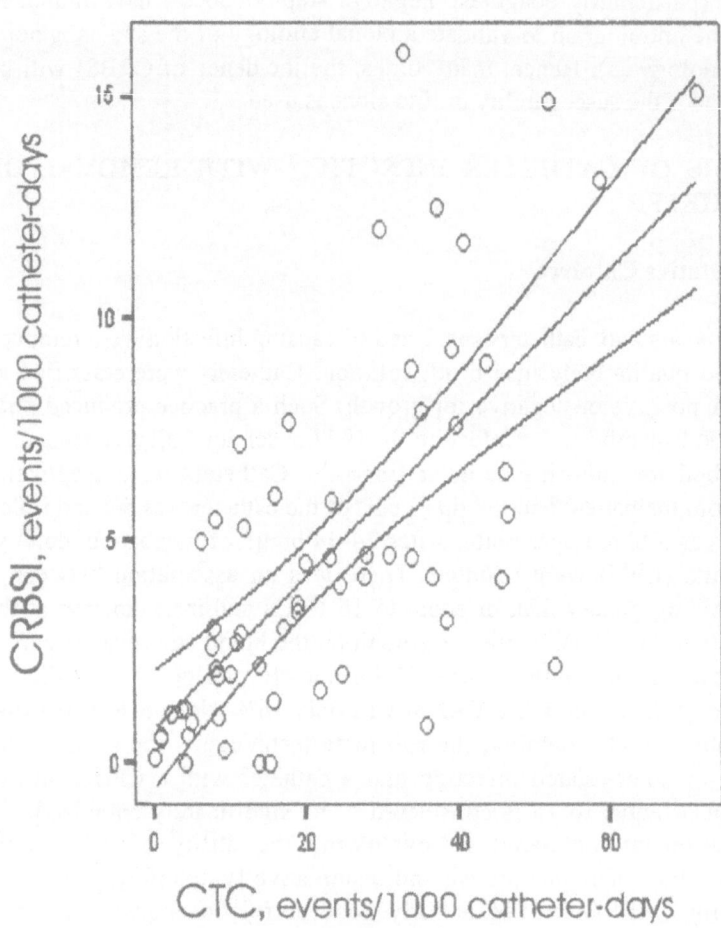

Figure 1. Relationship between catheter-related bloodstream infection (CRBSI) and catheter-tip colonization (CTC) utilizing data collated from 29 prospective studies on catheter-related infection. (From Rijnders, et al. (5) with permission)

Quantitative Cultures

During the 1980's, investigations revealed that catheter-related infection due to intraluminal rather than external colonization occurred much more frequently than had been previously reported (6,7). Since the roll-plate method of culturing samples only the external surface of the catheter, instances of catheter-related infection or CRBSI due to luminal bacterial colonization might be missed. Several culturing techniques have been developed to assay both the external and luminal surfaces.

One method involves immersing the suspected catheter in trypticase soy broth and flushing the lumen with broth (8). The broth media is serially diluted, inoculated onto blood agar plates and incubated for 72 hours at 37oC. All patients who had CRBSI grew > 1000 CFU; thus, that density of growth was defined as the threshold for diagnosing catheter infection.

Another methodology developed (9,10) was a quantitative technique in which a catheter tip is immersed in trypticase soy broth and sonicated in a water bath for one minute at 55,000 Hz. The broth is serially diluted onto blood agar plates and cultured for 48 hours. This technique also assays both the external surface and lumen of the catheter. CFUs of greater than or equal to 1000 are associated with CRBSI and catheter-related infection. This method has a higher sensitivity and an equivalent specificity for CRBSI when compared to the roll plate method (11).

Special consideration must be given to antiseptic or antibiotic-impregnated catheters. Several investigations have shown that a significant amount of chlorhexidine and silver sulfadiazine is found in broth following sonication of the catheter. In addition, other studies have shown that the antimicrobial compounds readily elute from the catheter after sonication and during culture in the blood agar plate (12). Consequently, the bacterial load on the catheter may be underestimated. Use of antiseptic "neutralizers" may improve the diagnostic yield when using the sonication culturing method (13). Such an option is not available for antibiotic-impregnated catheters, but an ultracentrifugation and decantation method has been described which improved the accuracy of the sonication culture results (14).

Catheter Staining

Because the results of the catheter cultures take several days, interest developed in the catheter stains to diagnose infection.

One group of investigators (15) described a Gram stain technique of catheters that allowed rapid diagnosis of catheter infection. Upon removal, catheters were immersed in gentian violet for 10 seconds, washed with tap water, immersed in Gram's iodine for 10 seconds, washed, decolorized with 95% ethanol, washed, immersed in safranin for 10 seconds and washed again. The catheter was then microscopically examined under oil-immersion fields (200). A typical exam took 15 minutes. The authors defined a positive finding as 1 organism discovered per 20 oil-immersion fields. Compared to the roll plate method, this technique resulted in a sensitivity and specificity of 100% and 96.9%, respectively.

Acridine orange (AO) binds to bacterial nucleic acid and stains the bacteria orange. This dye has been used as means to diagnose catheter infection (16). After catheter removal and fixing the catheters at 56oC for 2 minutes, the catheters were immersed in AO for 3.5 minutes, washed in water and air dried. If the devices failed to fluoresce under x100 magnification, they were considered negative. If fluorescence was present, magnification was increased to 1000 (oil-immersion) and the presence of organisms was determined. Compared to the roll plate method, the AO technique demonstrated a sensitivity of 84% and a specificity of 99%. AO detected all cases of CRBSI and the negative predictive value of the staining was 99%.

Despite the availability of these techniques to diagnose catheter infection more rapidly than the traditional culture techniques, neither of these methods has gained wide-spread popularity, primarily because of the time required by the microbiology technician to perform the staining and microscopic examination (17). In addition, a subsequent study casts doubt on the utility of either of these methods to diagnose infection (18).

Diagnosis of Infection with the Catheter *in situ*

A more valuable diagnostic test would be one in which catheter infection could be diagnosed with the catheter in place thereby obviating the need to remove catheters that are not the cause of infection.

Insertion Site

Clearly, if there is purulence at the exit site of the catheter, infection is present and the catheter should be removed. With signs of inflammation such as erythema, tenderness or warmth, catheter infection is likely and the catheter should also be removed. This caveat must be tempered with the realization that some catheter coatings may be associated with a higher risk of phlebitis (19). The lack of inflammation is not a reliable indicator of the absence of infection particularly in regard to infection with relatively avirulent organisms such as coagulase negative staphylococci (20).

Quantitative Blood Cultures

Although the data are conflicting, qualitative blood cultures drawn through catheters generally are confusing (if positive) and should not be performed unless a peripheral venipuncture is impossible (21). A number of studies have been published that have examined the utility of quantitative blood cultures drawn through the catheter with or without comparison to peripheral quantitative blood cultures. This method assumes that a large reservoir of bacteria resides in the lumen of the infected catheter and quantitative cultures will provide a reliable assay of the density of bacteria. Generally, these methods require use of a sterile non-bacteriostatic anticoagulant. The culture itself is most often accomplished using a pour plate technique although some investigators have inoculated chocolate blood agar plates with 0.5 ml of blood right at the bedside of the patient. One study (22) reported that quantitative blood cultures (cutoff of 15 CFU bacteria) drawn through a catheter suspected of causing infection had a sensitivity of 100% and a specificity of 94% when compared to the roll plate method. In addition, the negative predictive value was 100% but the positive predictive value was only 60%. Other investigations have examined the differential in CFU counts between quantitative catheter and peripheral blood cultures. In one study, a ratio of greater than 7 (catheter blood/peripheral blood) achieved a sensitivity of 78% and a specificity of 100% to detect CRBSI23. Other studies have reported similar results. These techniques have not assumed widespread clinical use because of the time, cost and difficulty required to perform the cultures.

Quantitative Skin Cultures

One of the primary mechanisms by which catheters become colonized and ultimately cause infection is growth of bacteria on the skin and propagation along the subcutaneous tract traversed by the catheter. As a consequence, there may be a link between the density of microbial growth at the insertion site and catheter-related infection. Some studies have shown that quantitative skin cultures are of value in either predicting or excluding catheter-related infection. However, in depth investigations using molecular epidemiology have shown that skin colonization with bacteria is a dynamic process and the correlation between skin organisms and bacteria on the tip of the catheter is poor. Using pulsed-field gel electrophoresis to identify accurately individual strains of bacteria, a positive predictive value of approximately 19% for quantitative skin cultures to predict catheter-related infection was reported24. A negative culture may be of value in excluding catheter-related infection in that the negative predictive value exceeded 80%.

Differential Time to Positivity

Automated blood culturing instruments sample media for bacterial growth throughout the day. When bacterial growth reaches a threshold density based on the level of fluorescence, carbon dioxide production or pH, the instrument reports a growth index and the culture is considered "positive". The media is removed for subculture and Gram stain. Rogers and Oppenheim were one of the first to evaluate the link between microbial inoculum and the time to the determination of a positive blood culture ("positivity") (25). They found in an in vitro study that the time to positivity of a blood culture was strongly correlated with the initial bacterial inoculum. There was an average decrease of 90 minutes for a blood culture to register as positive for each 10 fold increase in the bacterial concentration. Independently, another study showed a logarithmic relationship between the initial inoculum and time to positivity of the blood culture (Figure 2). These investigators reasoned that a blood culture drawn through a catheter with a large inoculum of bacteria in the lumen would become positive more quickly than a blood culture obtained from a peripheral vein (26). Consequently, they evaluated 64 patients with suspected CRBSI and performed paired blood cultures. The differential time to positivity was substantially greater in patients who had documented CRBSI. A cutoff of 120 minutes (time between the catheter blood culture

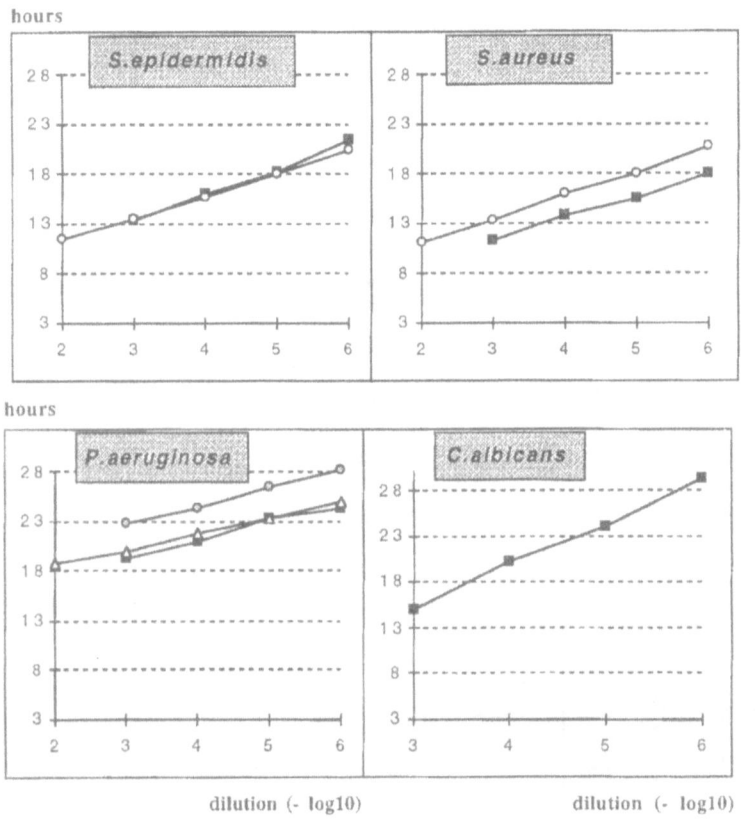

Figure 2. Relationship between initial microbial inoculum and time for the blood culture to become positive. (From Blot, et al. (26) with permission)

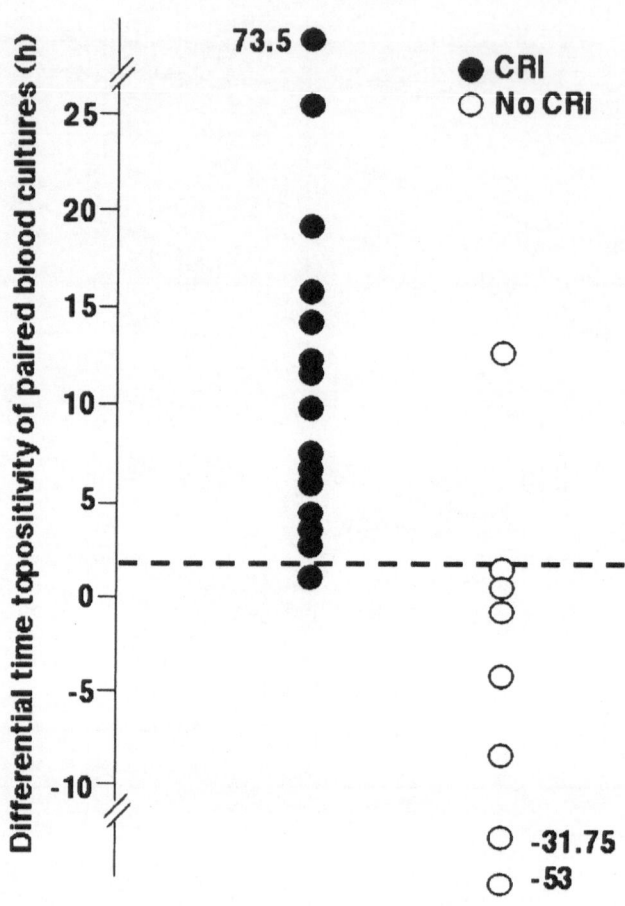

Figure 3. Differential time to positivity of paired blood cultures in patients with catheter-related sepsis, infection of other causes and indeterminate diagnoses. (From Blot, et al. (26) with permission)

turning positive and the peripheral blood culture turning positive) showed a sensitivity of 96% and a specificity of 100% for the diagnosis of CRBSI (Figure 3). This analysis excluded 22 patients in whom the diagnosis of CRBSI was uncertain because only one of the two blood cultures was positive. In a follow up study, 93 catheters were removed because of the suspicion of catheter-related infection. Using a cut-off of 120 minutes, these investigators demonstrated similar findings for paired blood cultures that were positive (27). Nineteen catheters had discordant cultures: either the catheter blood culture was positive and the peripheral blood culture was negative or vice versa. Of these, there were 3 confirmed cases of catheter-related infection. More recently, another group of investigators studied 21 patients with suspected catheter infection (28). They used receiver operating characteristic (ROC) curves to determine the optimal threshold for the differential time point and determined that the optimal cutoff point was 3 hours (rather than 2). The specificity was 100% and sensitivity was 81%. It is noteworthy that these three studies were all from cancer referral centers where a significant proportion of catheters were tunneled and long term: a situation where endoluminal colonization assumes more importance in the pathogenesis of catheter-related infection. Results from a different (albeit small) study where catheters were removed from patients with suspected catheter-related infection hospitalized in a mixed medical/surgical ICU failed to show a benefit of the differential time to positivity (DTTP) test (29). The sensitivity was only 25% and the positive predictive value was 33%. The implication from this study is that for short term catheters, DTTP may be less valuable because most infections in this population occur via the subcutaneous route. However, there are abstracted data which cast doubt on this conclusion (30). Clearly, more data from larger and more diverse patient populations are needed before the role of DTTP in the in situ diagnosis of CRBSI can be determined with certainty.

Endoluminal Brush

The endoluminal brush is a sterile nylon bristled tapered brush (8 mm long) attached to a stainless steel wire and enclosed in a polythene sleeve (Figure 4). The brush is introduced into one of the hubs of the catheter and advanced to the tip of the catheter. After removal, the brush is cut sterilely and placed in a sterile container with 1 ml of phosphate buffered saline (PBS). The container is sonicated and vortexed. The PBS is inoculated onto blood agar plates and cultured. Significant growth is considered to be > 1000 CFUs.

In a study involving 230 central venous catheters, endoluminal brushing demonstrated a sensitivity of 95% and a specificity of 84% for the diagnosis of CRBSI compared to a sensitivity of 82% and a specificity of 66% for the roll plate method (31). By contrast, a different group of investigators failed to show in a smaller study that this sampling brush was of benefit in the in situ diagnosis of catheter-related infection (32). Of note, however, these investigators did not sonicate the brush before quantitative cultures were performed. Such a difference in culture methods could explain the discordant results. Finally, there are some data suggesting that if the endoluminal brush is extended to the tip of the catheter, bacteria can be expelled into the circulation. If the brush is not advanced within 2 cm of the catheter tip and blood is aspirated from the lumen following brushing, a bacteremia is unlikely to be induced (33).

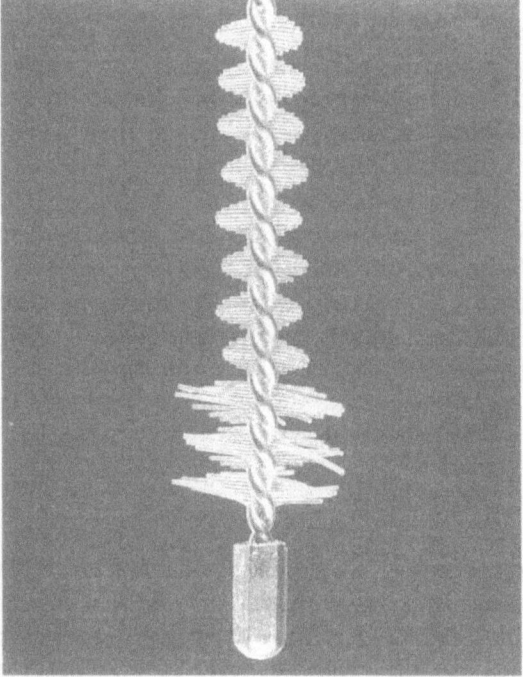

Figure 4. The endoluminal brush. (From van Heerden, et al.[32] with permission).

Acridine-Orange Leukocyte Cytospin (AOLC) Test

The AOLC test is another means of diagnosing catheter-related infection. In this technique, blood is aspirated and treated with edetic acid (EDTA). The sample is mixed with a formalin-saline mixture and centrifuged. The pellet is vortexed and transferred to a cytospin cupule and centrifuged again in a cytocentrifuge. The resulting pellet is placed on a microscope slide, heat dried and then stained with acridine orange. After examination under at least 100 high powered fields, the presence of at least one microorganism is considered positive. In a study of 128 cases of suspected CRBSI, this method compared favorably with the roll plate methods and endoluminal brush methods (34). In contrast, others (35) have not found the AOLC test sensitive enough to detect catheter-related sepsis; however, when it is combined with endoluminal brushing, the sensitivity improves significantly. Although this method appears to be at least as accurate as the "gold standard" for diagnosing CRBSI and can diagnose infection with the catheter in place, it is labor intensive and for this reason may not supplant other methods of diagnosing infection.

Serological Tests

Some of the most common organisms causing CRBSI include *Staphylococcus aureus* and *Staphylococcus epidermidis*. Both of these bacteria produce an extracellular material that has been identified as a short-chain length form of lipoteichoic acid (36). Patients develop IgM and IgG antibodies to this product; hence, these antibodies may serve as a serological marker for catheter infection caused by these bacteria.

An enzyme-linked immunosorbent assay (ELISA) for these antibodies has been developed (36) and used to determine whether it would have detected CRBSI in a group of patients with known catheter-related sepsis. The controls were a group of patients with central venous catheters without any evidence of sepsis. There were significant differences in mean IgG and IgM titers between the two groups but there was significant overlap in IgM titers (Figure 5). The IgG titer had sensitivity of 75% and specificity of 90% for the diagnosis of catheter-related sepsis compared to traditional clinical and microbiological criteria. Further study will be required to determine the role, if any, for this novel serological test in the diagnosis of catheter-related infection.

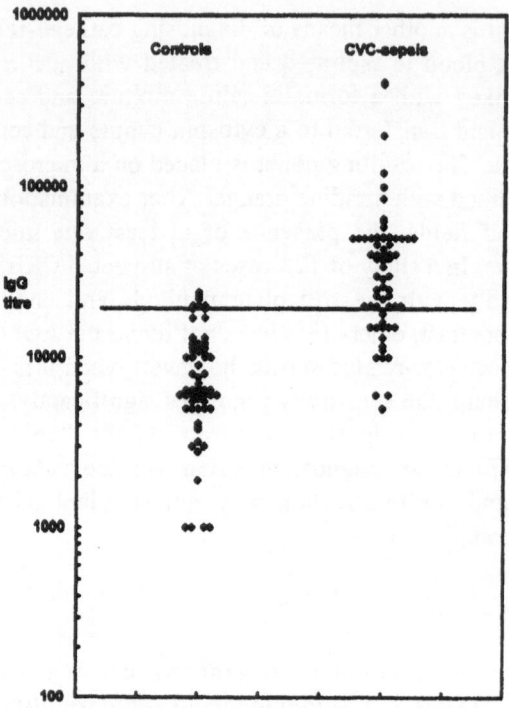

Figure 5. Scatter plots of IgG titers from patients with catheter-related sepsis and control patients. Points above the line (20,000 cut-off point) are considered a positive test. (From Elliott, et al.36 with permission).

CONCLUSIONS

Despite recent advances in the diagnosis of catheter-related infection with the catheter in situ, confirmation of the diagnosis still requires either the removal of the catheter and clinical improvement of the patient with signs of sepsis (e.g. defervescence and decreasing white blood cell count) or cultures from the removed catheter and peripheral blood yielding the same organism. Of the recent efforts to confirm the diagnosis of infection with the catheter in place, differential time to positivity remains the most simple and cost effective. However, further data will be required to define the utility of this method in the diagnosis of catheter-related infection and CRBSI.

REFERENCES

1.   O'Grady NP, Alexander M, Dellinger EP, Gerberding JL, Heard SO, Maki DG, Masur H, McCormick RD, Mermel LA, Pearson ML, Raad II, Randolph A, Weinstein RA. Guidelines for the prevention of intravascular catheter-related infections. Pediatrics 2002; 110(5):e51.
2.   Mermel LA. Defining intravascular catheter-related infections: a plea for uniformity. Nutrition 1997; 13:2S.
3.   Dobbins BM, Kite P, Kindon A, McMahon MJ, and Wilcox MH. DNA fingerprinting analysis of coagulase negative staphylococci implicated in catheter related bloodstream infections. J Clin Pathol 2002;55:824.
4.   Maki DG, Weise CE, and Sarafin HW. A semiquantitative culture method for identifying intravenous-catheter- related infection. N Engl J Med 1977;296:1305.
5.   Rijnders BJ, Van Wijngaerden E, and Peetermans WE. Catheter-tip colonization as a surrogate end point in clinical studies on catheter-related bloodstream infection: how strong is the evidence? Clin Infect Dis 2002;35:1053.
6.   Linares J, Sitges-Serra A, Garau J, Perez JL, and Martin R. Pathogenesis of catheter sepsis: a prospective study with quantitative and semiquantitative cultures of catheter hub and segments. J Clin Microbiol 1985;21:357.
7.   Sitges-Serra A, Puig P, Linares J, Perez JL, Farrero N, Jaurrieta E, and Garau J. Hub colonization as the initial step in an outbreak of catheter-related sepsis due to coagulase negative staphylococci during parenteral nutrition. JPEN J Parenter Enteral Nutr 1984;8:668.
8.   Cleri DJ, Corrado ML, and Seligman SJ. Quantitative culture of intravenous catheters and other intravascular inserts. J Infect Dis 1980;141:781.
9.   Sherertz RJ, Raad II A, Belani LC, Koo, KH, Rand DL Pickett SA. Straub, and. Fauerbach LL Three-year experience with sonicated vascular catheter cultures in a clinical microbiology laboratory. J Clin Microbiol 1990;28:76.
10.   Heard SO, RF Davis RJ, Sherertz MS, Mikhail RC, Gallagher AJ, Layon, and Gallagher T. J. Influence of sterile protective sleeves on the sterility of pulmonary artery catheters. Crit Care Med 1987;15:499.
11.   Sherertz RJ, Heard SO, and Raad II IA. Diagnosis of triple-lumen catheter infection: comparison of roll plate, sonication, and flushing methodologies. J Clin Microbiol 1997;35:641.
12.   Schierholz JM, Bach A, Fleck C, Beuth J, Konig D, and Pulverer G. Measurement of ultrasonic-induced chlorhexidine liberation: correlation of the activity of chlorhexidine-silver-sulfadiazine-impregnated catheters to agar roll technique and broth culture. J Hosp Infect 2000;44:141.
13.   Schmitt SK, Knapp C, Hall GS, Longworth DL, McMahon JT, and Washington JA. Impact of chlorhexidine-silver sulfadiazine-impregnated central venous catheters on *in vitro* quantitation of catheter-associated bacteria. J Clin Microbiol 1996;34:508.

14. Schierholz JM, and Pulverer G. Development of a new CSF-shunt with sustained release of an antimicrobial broad-spectrum combination. Zentralbl Bakteriol 1997;286:107.

15. Cooper GL, and Hopkins CC. Rapid diagnosis of intravascular catheter-associated infection by direct Gram staining of catheter segments. N Engl J Med 1985;312:1142.

16. Zufferey J, Rime B, Francioli P, and Bille J. Simple method for rapid diagnosis of catheter-associated infection by direct acridine orange staining of catheter tips. J Clin Microbiol 1988;26:175.

17. Braunstein H. Rapid diagnosis of intravascular catheter-associated infection by direct Gram staining of catheter segments. N Engl J Med 1985;313:754.

18. Coutlee F, Lemieux C, and Paradis JF. Value of direct catheter staining in the diagnosis of intravascular-catheter-related infection. J Clin Microbiol 1988;26:1088.

19. Carruth WA, Byron MP, Solomon DD, White WL, Stoddard GJ, Marosok RD, and Sherertz R. J. Subcutaneous, catheter-related inflammation in a rabbit model correlates with peripheral vein phlebitis in human volunteers. J Biomed Mater Res 1994;28:259.

20. Safdar N, and Maki DG. Inflammation at the insertion site is not predictive of catheter-related bloodstream infection with short-term, noncuffed central venous catheters. Crit Care Med 2002;30:2632.

21. Juste RN, Hannan M, Glendenning A, Azadian B, and Soni N. Central venous blood culture: a useful test for catheter colonisation? Intensive Care Med 2000;26:1373.

22. Vanhuynegem L, Parmentier P, and Potvliege C. 1988. In situ bacteriologic diagnosis of total parenteral nutrition catheter infection. Surgery 1988;103:174.

23. Fan ST, Teoh-Chan CH, and Lau KF. Evaluation of central venous catheter sepsis by differential quantitative blood culture. Eur J Clin Microbiol Infect Dis 1989;8:142.

24. Atela I, Coll P, Rello J, Quintana E, Barrio J, March F, Sanchez F, Barraquer P, Ballus J, Cotura A, and Prats G. Serial surveillance cultures of skin and catheter hub specimens from critically ill patients with central venous catheters: molecular epidemiology of infection and implications for clinical management and research. J Clin Microbiol 1997;35:1784.

25. Rogers MS, and Oppenheim BA. The use of continuous monitoring blood culture systems in the diagnosis of catheter related sepsis. J Clin Pathol 1998;51:635.

26. Blot F, Schmidt E, Nitenberg G, Tancrede C, Leclercq B, Laplanche A, and Andremont A. Earlier positivity of central-venous- versus peripheral-blood cultures is highly predictive of catheter-related sepsis. J Clin Microbiol 36:105.

27. Blot F, Nitenberg G, Chachaty E, Raynard B, Germann N, Antoun S, Laplanche A, Brun-Buisson C, and Tancrede C. Diagnosis of catheter-related bacteraemia: a prospective comparison of the time to positivity of hub-blood versus peripheral-blood cultures. Lancet 1999;354:1071.

28. Malgrange VB, Escande MC, and Theobald S. Validity of earlier positivity of central venous blood cultures in comparison with peripheral blood cultures for diagnosing catheter-related bacteremia in cancer patients. J Clin Microbiol 2001;39:274.

29. Rijnders BJ, Verwaest C, Peetermans WE, Wilmer A, Vandecasteele S, Van Eldere J, and Van Wijngaerden E. Difference in time to positivity of hub-blood versus nonhub-

blood cultures is not useful for the diagnosis of catheter-related bloodstream infection in critically ill patients. Crit Care Med 2001;29:1399.

30.   Raad I, Hanna H, Alakech B, and et al. Diagnosis of catheter-related bloodstream infection: Differential time to positivity for short-term and long-term central venous catheters. In 40th Interscience Conference on Antimicrobial Agents and Chemotherapy, Toronto, p. 421.

31.   Kite P, Dobbins BM, Wilcox MH, Fawley WN, Kindon AJ, Thomas D, Tighe MJ, and McMahon MJ. Evaluation of a novel endoluminal brush method for in situ diagnosis of catheter related sepsis. J Clin Pathol 1997;50:278.

32.   van Heerden PV, Webb SA, Fong S, Golledge C L, Roberts BL, and Thompson WR. Central venous catheters revisited--infection rates and an assessment of the new Fibrin Analysing System brush. Anaesth Intensive Care 1996;24:330.

33.   Worthington T, Perry A, Lambert P, and Elliott T. Is brushing of central venous catheters safe clinical practice? J Hosp Infect 2002;51:149.

34.   Kite P, Dobbins BM, Wilcox MH, and McMahon MJ. Rapid diagnosis of central-venous-catheter-related bloodstream infection without catheter removal. Lancet 1999; 354:1504.

35.   Tighe MJ, Kite P, Thomas D, Fawley WN, and McMahon MJ. Rapid diagnosis of catheter-related sepsis using the acridine orange leukocyte cytospin test and an endoluminal brush. JPEN J Parenter Enteral Nutr 1996;20:215.

36.   Elliott TS, Tebbs SE, Moss HA, Worthington T, Spare MK, Faroqui MH, and Lambert PA. A novel serological test for the diagnosis of central venous catheter-associated sepsis. J Infect 2000;40:262.

Chapter 4

DIAGNOSIS OF CATHETER-RELATED INFECTIONS

Gérard Nitenberg, M.D., François Blot, M.D.
Service de Réanimation Polyvalente, Institut Gustave Roussy
Villejuif, France

Introduction

Intravascular catheters, mainly central venous catheters (CVCs), are widely used in the management of critically ill patients who are thereby exposed to intravascular catheter-related infections (CRIs) resulting in increased hospital costs, duration of hospitalization, and patient morbidity (1).

Clinical findings are not sufficient to establish the diagnosis of CRIs (2), which relies on microbiological criteria. Following a brief overview of the clinical pictures leading to the suspicion of CRI, we will focus on the microbiological tools available to establish the diagnosis, with or without catheter removal.

PATHOGENESIS

The two major pathways (extraluminal and intraluminal routes) involved in the colonization of the catheter occur at different times during catheterization (3). Overall, about 65% of CRIs originate from the skin, 30% from the contaminated hub, and 5% from other pathways (3). The extraluminal route (from the cutaneous entry site) predominates for short-term catheters, such as those inserted in the intensive care unit (ICU). Endoluminal contamination is the predominant portal of entry for long-term catheters, such as in total parenteral nutrition or in cancer patients (4,5). Although CRIs will be better diagnosed in critically ill patients by techniques exploring the extraluminal pathway of colonization, ideally, methods exploring both the external and internal surfaces of the catheter tip would be appropriate, whatever the mechanism involved.

DEFINITIONS AND CLINICAL DIAGNOSIS

Localized infections (confined to the catheter and surrounding tissues), and systemic infections (all types of bloodstream infections = CR-BSIs) must be considered separately.

Apart in some particular cases, clinical findings lack sensitivity or specificity to establish a definite diagnosis of CRI. In particular, local inflammation is not significantly associated with CRI, with the exception of gross pus at the catheter-exit site. In a recent study, the rate of colonized CVCs was similar whatever the indication for catheter removal: systemic sepsis (36.7%), local sepsis (36.4%), CVC no longer required (31.3%) or death (30%) (6).

Local Infections

Exit-site infections are defined by erythema, local warmth, tenderness, induration, or purulence within 2 cm of the catheter exit site (1). In tunnel infections these signs overly the catheter and are >2 cm from the exit site (1).

Inflammation at the catheter exit site is a poor indicator of the diagnosis of CRI (7). Purulence around the CVC in the presence of bacteremia is highly specific but poorly sensitive.1 In addition, there is no significant association between CRI and the presence of swelling, tenderness, or extravasation of fluid at the insertion site (8).

Systemic Infections

Catheter-related bloodstream infections (CR-BSI) are defined by the isolation of the same organism from a catheter segment culture and from the blood of a patient with accompanying clinical symptoms of bloodstream infection (BSI), in the absence of other apparent sources of infection.1 Infusate-related BSIs are defined by isolation of the same microorganism from infusate and from separate percutaneous blood cultures, without other identifiable sources of infection (1). Finally, when CR-BSI fails to resolve upon catheter removal and despite appropriate antimicrobial therapy, a suppurative thrombophlebitis should be suspected.

Fever, with or without chills, is the most sensitive finding (but has poor specificity) for the diagnosis of CR-BSI (1) especially in critically ill patients. In the absence of laboratory confirmation, the disappearance of the sepsis syndrome or return to normal body temperature after catheter removal may be considered indirect evidence of CR-BSI (1,9) Finally, a CR-BSI is likely when blood cultures grow a common skin organism (such as coagulase-negative staphylococci, Propionibacterium spp., micrococcus, or Bacillus spp.), Staphylococcus aureus or Candida spp., and no apparent source of sepsis other than the catheter is identified (9).

CATHETER-TIP CULTURE TECHNIQUES

Qualitative Broth Culture

A definite diagnosis of CRI traditionally requires the removal of the catheter or guidewire exchange for culture of the catheter tip.

Qualitative broth cultures of the catheter tip by immersion in liquid medium are highly sensitive but poorly specific, and do not allow to distinguish between colonization and infection (3). The effect of contamination of the skin exit site and the subcutaneous tunnel on the accuracy of CVC tip cultures has been studied in an *in vitro* model: (10) pulling a catheter through a contaminated area results in catheter-tip contamination, and organisms can be dislodged from the surface of the distal segment of the catheter when it is pulled through an agar tunnel.

Semi-quantitative Catheter-Tip Culture

The semi-quantitative roll plate method, popularized by Maki *et al.* (7), allows to distinguish infection from contamination, and is highly efficient in diagnosing CRI. The external surface of the distal segment (4-5 cm) of the catheter is rolled on a blood agar plate and subsequently incubated at 37°C for 48 hours. A threshold of 15 colony forming units (cfu) is significantly correlated with local signs of inflammation (Table 1).

This technique, which explores only the external surface of the catheter, has some important limitations. First, the threshold of 15 cfu for positivity of the semi-quantitative culture was built on the presence or absence of local inflammation (7), but correlation with clinical criteria of infection was not indicated in the study. Second, as the study included a large majority of peripheral short-term catheters, the validity of extrapolation to CVCs (especially long-term intravascular devices for which intraluminal colonization predominates) remains to be proven. Third, the technique lacks specificity (20-50%) (3).

Table 1. Catheter-tip Culture Methods.

	N°	Setting	Method	CVC/periph.	CRIs (%)	Mean duration of placement	Threshold (cfu/ml)
Maki (7)	250	All hosp	Roll plate	CVC 33 Periph. 217	10	3 days	15
Brun-Buisson (16)	331	ICU	Vortexing	CVC only	11	5 days	1000
Sherertz (17)	1681	All hosp	Sonication	CVC + periph.	12	NA	100

N°: number of catheters studied
ICU: intensive care unit
CVC: central venous catheter
Periph: peripheral catheter
CRI: catheter-related infection
Cfu/ml: colony forming units per milliliter
NA: not available

For these reasons, other thresholds for significant colonization of the catheter have been evaluated, but with disappointing results. In the study of Collignon *et al.* on 780 CVCs inserted in 440 critically ill patients, the sensitivity and specificity of a threshold of 5 cfu were 92% and 83% respectively, with a negative predictive value of 99.8%. The positive predictive value of only 8.8%, linked to the low incidence of CR-BSIs (2%), was not significantly improved (9.8%) when the threshold was increased to 100 cfu (11). Kristinsson *et al.* studied prospectively 236 CVCs using roll-plate culture, combined with tip-flush and ultrasonication techniques (12).

Although the negative predictive values were high (99% and 93% for thresholds of 15 and 100 cfu, respectively), the positive predictive values remained low, 46% and 56%, respectively). Similar findings were reported by Rello *et al.* in critically ill patients using thresholds of 15 and 50 cfu (13).

Interestingly, a study of 197 CVCs comparing the value of semi-quantitative tip cultures plated at the bedside with those cultured in the laboratory suggested that bedside plating could be more sensitive than roll plate cultures performed in the laboratory (14). It is conceivable that organisms such as Candida spp., known to thrive in moist environments, may not survive for a prolonged period on dry plastic surfaces.

In summary, although physicians have to be aware of the limitations of the roll-plate culture, the semi-quantitative culture technique is considered as easy and fast, and remains the most commonly used method world-wide for the diagnosis of significant catheter-tip colonization.

Quantitative Catheter-Tip Culture Techniques

The quantitative tip-flush technique proposed by Cleri *et al.* permits to explore only the internal part of the CVC (15) and demonstrated that when using a threshold for positivity of 103 cfu/ml, a good correlation with catheter-related bacteremias was shown.

However, the method is cumbersome and not always easy to perform. Therefore, a simplification of the technique for routine clinical practice using catheter vortexing in sterile water was proposed by Brun-Buisson et al, with the advantage of exploring both the external and internal surface of the catheter (16) A threshold of 103 cfu/ml was correlated with systemic signs of infection, with or without catheter-associated bacteremia, and exhibited high specificity (88%) and sensitivity (97.5%) in critically ill patients with a CVC in place for several days (Table 1).

Another technique for quantitative catheter-tip culture using ultrasonication to dislodge bacteria adherent to the catheter gave similar results (17). The sonication method allowed quantitation of the number of cfu removed from a catheter of between 102 and 107 cfu. For catheter cultures in which $\geq$ 102 cfu grew, the likelihood of positive blood cultures for the same organism increased with the number of organisms recovered from the catheter (Table 1). Disadvantages of the method include the need for additional equipment, and the difficult standardization of the ultrasound technique.

Comparison of Catheter Culture Techniques

The different mechanisms of colonization of the intravascular part of the catheter may explain some discrepancies between the catheter-tip culture techniques. The optimal technique to determinine that the catheter is the source of BSI should be independent of the duration of placement and the route of colonization.

Only quantitative or semi-quantitative cultures of catheters are recommended by the Guidelines for the Management of Intravascular CRI.1 Sensitivity and specificity are dependent on the definition used for the diagnosis of CR-BSI, which may result in potential misclassification bias (18) Receiver operating characteristic (ROC) analysis has been used to determine which culture technique proposed for the diagnosis of CR-BSI offers the best overall performance. In a meta-analysis focusing on the diagnostic tools for CRI (18) it has been suggested that the diagnostic accuracy increased with better quantitation (i.e., quantitative > semi-quantitative > qualitative methods) (19). In a recent study, Kite *et al.* reported that the specificity of the roll tip method was lower (55%) than the tip flush (76%) and the endoluminal brush (98%) techniques, whereas the sensitivity of all methods was greater than 90% (20). When several culture techniques were compared for the diagnosis of triple-lumen CRI, sonication was 20% more sensitive than the roll plate method (21). In conclusion, the vortexing or ultrasonication techniques, which take into account both the external and internal surfaces of the device, presently represent the optimal methods for the diagnosis of CRI, whatever the route of colonization and the duration of placement.

Diagnosis of CRI in Pulmonary Artery Catheters

To diagnose pulmonary artery CRIs, the need to culture the distal segments of both the pulmonary artery catheter and the indwelling introducer has been emphasized (22). Recently, we reported that colonization of the introducer and the Swan-Ganz catheter were dissociated in 6 of 7 episodes of CRI (23). Introducers were mainly colonized during the first 5 days, while Swan-Ganz catheter colonization occurred later, thus suggesting that extraluminal contamination of introducers occurs early from the skin, whereas Swan Ganz contamination results from endoluminal contamination resulting from repeated handling.

DIAGNOSIS OF CRI WITHOUT CATHETER REMOVAL

Only 15 to 25% of CVCs removed because of suspected infection actually prove to be infected upon quantitative catheter-tip cultures (2,9). Therefore, diagnostic techniques have been proposed to establish the diagnosis of CRI in situ and to avoid unjustified catheter removal and potential risks associated with the placement of a new catheter at a new site or through guide-wire exchange.

Culture of Entry Site and Catheter Hub

Culture of skin and entry site of the catheter, and cultures of the hub have high sensitivity and high negative predictive value: they are therefore mainly destined to rule out the diagnosis of CRI (5).

Cultures of the catheter entry site (Table 2) explore mainly the extraluminal contamination. In patients receiving total parenteral nutrition (24), the growth of >1000 cfu at the catheter site was significantly associated with CRI. Using a threshold of 15 cfu in critically ill patients, the method was considered useful for assessing catheter colonization, whatever the reason for CVC removal (25). Mahé *et al.* found that sensitivity, specificity, and positive and negative predictive values of skin culture for detection of CVC colonization in ICU patients were 92.3%, 52.7%, 32%, and 96.7%, respectively (26).

The values of targeted and surveillance skin cultures were compared in 132 cancer patients with long-term CVCs (27). Targeted skin cultures were associated with a sensitivity of 75%, a specificity of 100%, a positive predictive value of 100%, and a negative predictive value of 92%, whereas operational values for surveillance skin cultures remained very low. Therefore, targeted quantitative skin cultures (when CRI is suspected) are useful to rule out the diagnosis of CRI, and surveillance skin cultures are not recommended.

Table 2. Hub and Skin Exit-Site Culture Methods.

	Skin / Hub	N°	Setting	Suspicion of CRI / No suspicion	CVC/ peri.	CRIs (%)	Diagnostic criterion	Mean duration of placement	Threshold (cfu)	Se (%)	Sp (%)	NPV (%)	PPV (%)
Bjornson (24)	Skin	74	TPN		CVC	26	Roll plate		10^2 /25 cm^2	68	91	91	68
Cercenado (5)	Skin +Hub	139	All hosp.	Yes 79/No 60	CVC 95 Peri 44	38	Vortexing	15 (1-21) d 5 (1-15) d	15 / 10 cm^2	T:97	T.68	T:99 All:96	T 34 All-66
Armstrong (8)	Skin	152	TPN		CVC	13	Roll plate		50 / 5 cm^2	45	94	92	53
Gudet (25)	Skin	50	ICU	Yes 20/No 30	CVC	20	Vortexing	7 days	15 / 9 cm^2	100		100	
Gudet (25)	Hub	"	"	"	"	"	"	"	>0	45			
Raad (27)	Skin	132	Cancer	Yes 15/No 132	CVC	20	Roll plate + Sonication	> 3 months	10^3 /24 cm^2	T:75 S:18	T:100 S:92	T:92 S:93	T:100 S 25
Mahé (26)	Skin	134	ICU	Yes 60/No 74	CVC		Vortexing	10 days	15 /25 cm^2	92	52	96	32
Segura (31)	Skin +Hub	41	TPN	Yes	CVC	32	Roll plate	22 days	>0			96*	100*

N°. number of catheters studied
ICU: intensive care unit
TPN. total parenteral nutrition
CVC: central venous catheter
Peri: peripheral catheter
CRI: catheter-related infection
Cfu. colony forming units
T = targeted, S = surveillance cultures
* criterion: catheter-related bacteremia
Se: sensitivity
Sp: specificity
NPV: negative predictive value
PPV: positive predictive value

Cultures of the catheter hub (Table 2), which explore mainly the endoluminal mechanism, are more useful in patients with long-term catheters. Indeed, in 50 critically ill patients with a median duration of catheterization of 7 days (25), there was no case of catheter colonization with negative skin cultures and positive hub cultures. Similar findings were reported by Fortun *et al.* in 124 patients with nontunneled short-term CVCs (28). On the other hand, in patients on total parenteral nutrition, Sitges-Serra *et al.* showed that an infected hub was associated with an infected tip in 15 of 17 episodes of CRI due to coagulase negative staphylococci (29).

The respective predictive value for CR-BSI of hub and skin cultures was investigated in patients on total parenteral nutrition managed without removal of the central line (30) (Table 2). The negative predictive value of combined skin and hub cultures was 96%. A positive hub culture had a 100% positive predictive value for CR-BSI. In a population of inpatients housed in different hospital wards, Cercendao *et al.* (5) found that the predictive value of positive superficial cultures for the diagnosis of CRI was 66.2% and that of negative cultures, 96.7%. In the aforementioned study by Fortun *et al.* (28), the sensitivity of the combined skin and hub cultures increased to 86.2%. It is reasonable to conclude that in patients with suspected CRI but negative superficial cultures, the diagnosis of CRI may be ruled out (5,28,31).

Quantitative Blood Cultures

The aim of central venous (hub) quantitative blood cultures is to measure the number of microorganisms present in the blood drawn through the hub of the CVC. When a bacteremia is responsible for a CRI, the number of microorganisms retrieved by the hub blood culture is high, due to a purging effect of the infected lumen of the catheter. Pour plate cultures, lysis-centrifugation technique, and direct inoculation onto agar media are available for quantitative blood cultures. The lysis-centrifugation technique is effective in the rapid isolation of organisms from mixed cultures and is more sensitive than standard broth culture in detecting low inoculum size bloodstream infection caused by Enterobacteriaceae and yeasts, whereas contaminants are more frequent than in broth cultures (32).

Quantitative hub-blood cultures have been evaluated in several studies (33,34) (Table 3a). Using a cut-off value of > 100 cfu/ml with the pour plate technique, the sensitivity and specificity of quantitative hub-blood culture were 82% and 100%, respectively, in 64 patients with catheters remaining in situ for a mean of 19 (3-65) days (34). In 179 cancer patients, semi-quantitative hub-blood cultures (using a threshold of 103 cfu/ml) had a specificity of 99%, but a sensitivity of only 20% for catheter tip colonization (33). Quantitative central blood cultures are therefore characterized by high specificity and high positive predictive value, allowing to establish the diagnosis of CRI in case of positivity.

Table 3a. Quantitative Central Blood Culture Methods.

	N°	Setting	CRIs (%)	Culture method	Mean duration of placement	Threshold (cfu)	Se (%)	Sp (%)
Andremont (34)	205	Cancer	29	Pour plate		> 1000	20	99
Capdevila (35)	107	All hosp.	20	Pour plate	19 days	> 100	82	100

Paired Central and Peripheral Quantitative Blood Cultures

When a CRI is present, the comparison of the microbial count between simultaneous hub and peripheral blood cultures shows an overload of bacteria on the central blood culture, compared to the peripheral blood culture. In other cases of bloodstream infection, the microbial counts are similar.

The value of differential quantitative blood cultures has been assessed in several studies (Table 3b). A significant differential colony count of 4 to 10:1 for the CVC vs. the peripheral vein culture is indicative of CRI.34-37 Raucher *et al.* showed that a 10-fold or greater difference in bacterial concentrations between the two specimens was indicative of CR-BSI in children with Broviac catheters (37). Using a cut-off of 4:1, a sensitivity of 94%, with specificity and positive predictive values of 100% were obtained in patients hospitalized in different wards (35). Using pediatric Isolator® tubes in 58 bacteremic adult patients, a specificity and a positive predictive value of 100%, with slightly lower sensitivity (83%) and negative predictive value (78%) have been reported using a cut-off value of 3:1 (34). In practice, the differential colony count usually exceeds 50 or 100 in the case of proven CRI.

Table 3b. Paired Quantitative Central Blood Culture Methods.

	N°	Setting	CRIs (%)	Culture method	Threshold (ratio)	Se (%)	Sp (%)
Raucher (37)	28	Children	25	Direct inoculation	10 : 1	100	100
Capdevila (35)	107	All hosp.	20	Pour plate	4 : 1	94	100
Flynn (33)	13	Cancer + TPN	61	Isolator 1.5	5 : 1		100
Fan (32)	24	TPN	37	Pour plate	7 : 1	78	100
Douard (34)	58	Bacteremic adults	(all)	Isolator 1.5	3 : 1	83	100
Quilici (38)	283	ICU	19	Pour plate	8 : 1	93	99

N°: number of catheters studied
ICU: intensive care unit
TPN: total parenteral nutrition
CRI: catheter-related infection
Cfu: colony forming units
Se: sensitivity
Sp: specificity

Paired quantitative blood cultures have been validated recently for short-term catheters in the intensive care unit by Quilici *et al.* (38). ROC curve analysis was carried out by varying the catheter/peripheral cfu ratio. The combined sensitivity and specificity were most satisfactory at a ratio of 8. With the use of this threshold, differential blood cultures had a sensitivity of 92.8% and a specificity of 98.8%. The specificity was 100% when the analysis was restricted to catheters removed because of suspected CRI.

Despite the efficacy of the method, differential quantitative blood cultures are not routinely used in clinical practice, mainly because of their relative complexity and cost. To overcome these problems, new diagnostic tools have been recently developed.

Paired Central and Peripheral Non-Quantitative Blood Cultures

The measurement of the differential time to positivity between hub-blood and peripheral blood cultures has been proposed by our group (39). The time to blood culture positivity may be measured in clinical microbiology practice using automatic devices. A given cut-off value, linked to the metabolism and to the number of microorganisms initially present, indicates that bacterial or fungal growth has occurred in the bottle. The higher the initial bacterial inoculum, the quicker this cutoff value is reached. In *in vitro* study,40 a linear relation between the initial concentration of various microorganisms and the time to positivity has been shown for all species tested.

In patients with long-term catheters, an earlier positivity of central vs peripheral vein blood cultures was shown to be highly predictive of CRI. A cut-off limit of 2 h had sensitivity and specificity above 95% for the diagnosis of CRI (40). These results have been confirmed by a prospective study (39): a definite diagnosis of CRI could be made in 16 of the 17 patients with the same threshold for the differential time to positivity of 2 h; overall sensitivity was 91% and specificity, 94%. More recently, in another study in 107 cancer patients, a ROC curve was constructed to determine the optimum threshold of the test; a cut-off point of 3 h was associated with 100% specificity and 81% sensitivity (41). For an accurate interpretation of the differential time to positivity, a rigorous method is mandatory. The first mL drawn via the catheter should be used for culture and not discarded, only aerobic bottles are needed; for multiple lumen catheters, blood should be drawn from the distal port, which corresponds to the portion of the device cultured (39). The value of this technique is greatest among patients with long-term catheters, which are predominantly colonized by an endoluminal route, than among those with short-term CVCs such as critically ill patients; (42) however, methodological biases in the latter study preclude definitive conclusions (43). In this way, Seifert *et al.* showed recently that the differential time to positivity method compared favourably with paired quantitative blood cultures for the diagnosis of CR-BSIs in neutropenic patients with short-term non-tunnelled catheters (44). Specific studies are needed to evaluate the validity of the technique in critically ill patients with short-term catheters.

The practicability of the method may have some limitations. First, the technique implies that clinical microbiology laboratories use continually-monitored, blood-culture systems. Second, a 24-hour duty staff in the microbiology laboratory (including weekends) or the possibility to process blood cultures on the ward are rather uncommon.

RAPID DIAGNOSIS OF CRI BY USING DIRECT EXAMINATION

Considering that an overnight incubation is usually necessary for microbiological cultures of a catheter segment, exit-site or blood samplings, an early diagnostic and microbiological orientation could be useful, e.g., for patients with severe sepsis of unknown origin.

Acridine-Orange Leucocyte Cytospin Test

Direct examination of blood drawn from the catheter using acridine-orange leucocyte cytospin (AOLC) test is a rapid method proposed for the diagnosis of CRI. The AOLC test allows to detect bacteria from a small sample (50 μL) of blood aspirated from the catheter (45). The cytospin allows the production of a monolayer from the sample onto a slide. Acridine-orange is an intercalating agent used to stain DNA on the slide, which may then be examined using ultraviolet microscopy with oil immersion. In a population of infants with suspected catheter sepsis defined by quantitative blood cultures (46), the AOLC test was 87% sensitive and 94% specific for the diagnosis of CRI. The results are available in an hour.

However, the AOLC test may be less sensitive for the detection of CRI in adults, because of a lower bacterial inoculum during bloodstream infection in adults than in neonates. Tighe *et al.* reported a modification of the test using an endoluminal brush to release larger number of organisms from the inner part of the catheter. After the brush is used, fibrin and organisms are released from the wall of colonized catheters, subsequently aspirated and identified using the AOLC test (45). Two groups of 50 adult patients with suspected sepsis in the presence of a CVC were compared. In the first group, a blood sample was drawn from the catheter for the AOLC test; in the second, an endoluminal brush was used to "sweep" the catheter before blood sampling. Results of the AOLC test were compared with culture of the catheter tip. The test was positive in only 12% of the infected catheters in group 1, compared to 83% in group 2. Despite these encouraging results of endoluminal brushing, the theoretical risk of embolization or subsequent bacteremia should be considered.

More recently, Kite *et al.* used AOLC test and Gram stain for rapid diagnosis of CRI without catheter removal in 124 surgical adult patients. The Gram stain and AOLC test is simple, rapid (30 min), inexpensive, and requires two 50 mL samples of catheter blood treated with edetic acid, and the use of light and ultraviolet microscopy. A sensitivity of 96%, a specificity of 91%, a negative predictive value of 97%, and a positive predictive value of 91% were reported for the diagnosis of bacteremic CRI with both tests taken together (20). Gram stain and AOLC test had a threshold of 1000 micororganisms/mL of blood; considering that peripheral blood contained less than 250 cfu/ml, the technique is unlikely to detect bacteremia unrelated to the catheter. The operational values reported were similar to those obtained using the measurement of the differential time to positivity (39).

Interestingly, while the specificity of the method was high in both studies, the discrepancy is important for sensitivity between the Kite study and the former Tighe study, but without any clear explanation. Microbiological techniques are similar, and the use of Gram stain in the latter study cannot account for the difference if we agree that the concordance between Gram stain and AOLC test is excellent (20). Despite these concerns, this promising technique allows rapid diagnosis (< 1 h), is easy to perform and could be recommended to establish the diagnosis of CRI without catheter removal, and to guide early targeted antimicrobial therapy (or avoid unnecessary antibiotic use).

Gram Staining of Blood Drawn from the Catheter

Gram staining of blood drawn from the catheter is a simple method for the diagnosis of CRI, enabling a preliminary identification of the pathogen. Kite *et al.* have suggested that the AOLC test could be more accurate in case of Gram-negative bacteremia if bacterial counts are low and if red blood cells are poorly lysed (20). Nevertheless, the authors showed a high concordance between the results of Gram stain and the AOLC test. Using the Gram staining alone for direct examination of blood drawn from the catheter during 23 episodes of CRI, Moonens *et al.* reported a 100% specificity but a lower sensitivity (78% for definite CRI, 61% for suspected and definite CRI taken together) (47).

GUIDEWIRE EXCHANGE

A compromise solution between diagnostic techniques with and without removal of the catheter is guidewire exchange of CVCs suspected of CRI : if the first catheter is found significantly colonized, the second catheter is removed and a new line inserted in a new site. Considering that in about 80% of suspected CRI, the catheter is not the source of infection, guidewire exchange could prevent the numerous non-infectious complications associated with puncture at a new site (pneumothorax, hematoma, etc).

The value and safety of guidewire exchange in case of suspicion of CRI remain highly controversial. Although the technique is associated with fewer mechanical complications and less discomfort than new-site replacement, guidewire exchange could be linked to a slightly (non significant) greater risk of CRI (48). Therefore, if guidewire exchange is considered, meticulous aseptic technique is mandatory, and the method is precluded in case of signs of local inflammation or purulent discharge at the insertion site.

In the recent CDC Guidelines for the Prevention of Intravascular Catheter-Related Infections, a more stringent position on guidewire exchange was taken (49): "replacement of temporary catheters over a guidewire in the presence of bacteremia is not an acceptable replacement strategy, because the source of infection is usually colonization of the skin tract from the insertion site to the vein". Other experts consider this position somewhat excessive. In the recent revision of the XIIth Consensus Conference on Intravascular Catheter-Related Infections of the French Society of Critical Care (www.srlf.org), guidewire exchange was considered acceptable in case of low suspicion of CRI, in patients with strictly stable conditions, and without clinical signs of local inflammation.

CONCLUSIONS

New techniques have been recently proposed for the diagnosis of CRI. The most promising ones seem to be the direct examination of blood drawn from the catheter using the AOLC test and Gram stain, and the differential time to positivity of paired blood cultures. Both techniques could be more accurate for long-dwelling catheters, such as those used in cancer patients, than for short-term catheters. Therefore, these new tools have to be validated in different acute care settings before they can be recommended for routine use (50). The easiest method to set up immediately might be the differential time to positivity test, given many clinical microbiology laboratories use continuous monitoring blood culture systems, and many physicians investigate a new fever by drawing simultaneous catheter and venipuncture blood cultures

REFERENCES

1. Mermel LA, Farr BM, Sherertz RJ, *et al*. Guidelines for the management of intravascular catheter-related infections. Infect Control Hosp Epidemiol 2001; 22: 222-242.
2. Raad II, Sabbagh MF, Rand KH, Sherertz RJ. Quantitative tip culture methods and the diagnosis of central venous catheter-related infections. Diagn Microbiol Infect Dis 1992; 15: 13-20.
3. Blot F, Brun-Buisson C. Current approaches to the diagnosis and prevention of catheter-related infections. Curr Opinion Crit Care 1999; 5: 341-349.
4. Raad I, Costerton W, Sabharwal U, Sacilowski M, Anaissie F, Bodey GP. Ultrastructural analysis of indwelling vascular catheters: a quantitative relationship between luminal colonization and duration of placement. J Infect Dis 1993; 168: 400-407.
5. Cercenado E, Ena J, Rodriguez-Creixems M, Romero I, Bouza E. A conservative procedure for the diagnosis of catheter-related infections. Arch Intern Med 1990; 150: 1417-1420.
6. Juste RN, Hannan M, Glendenning A, Azadian B, Soni N. Central venous blood culture: a useful test for catheter colonisation ? Intensive Care Med 2000; 26: 1373-1375.
7. Maki DG, Weise CE, Sarafin HW. A semiquantitative culture method for identifying intravenous-catheter-related infection. N Engl J Med 1977; 296: 1305-1309.
8. Armstrong CW, Mayhall CG, Miller KB, Newsome HH, Sugerman HJ, Dalton HP, Hall GO, Hunsberger S. Clinical predictors of infection of central venous catheters used for total parenteral nutrition. Infect Control Hosp Epidemiol 1990; 11: 71-78.
9. Raad II, Bodey GP. Infectious complications of indwelling vascular catheters. Clin Infect Dis 1992; 15: 197-210.
10. Harris GJ, Rosenquist MD, Kealey GP. An *in vitro* model for studying the effect of the subcutaneous tunnel and the skin exit site on the accuracy of central venous catheter tip cultures. J Burn Care Rehabil 1992; 13: 628-631.
11. Collignon PJ, Soni N, Pearson IY, Woods WP, Munro R, Sorrell TC. Is semiquantitative culture of central vein catheter tips useful in the diagnosis of catheter-associated bacteremia ? J Clin Microbiol 1986; 24: 532-535.
12. Kristinsson KG, Burnett IA, Spencer RC. Evaluation of three methods for culturing long intravascular catheters. J Hosp Infect 1989; 14: 183-191.

13.   Rello J, Coll P, Prats G. Evaluation of culture techniques for diagnosis of catheter-related sepsis in critically ill patients. Eur J Clin Microbiol Infect Dis 1992; 11: 1192-1193.

14.   Hnatiuk OW, Pike J, Stoltzfus D, Lane W. Value of bedside plating of semiquantitative cultures for diagnosis of central venous catheter-related infections in ICU patients. Chest 1993; 103: 896-899.

15.   Cleri DJ, Corrado ML, Seligman SJ. Quantitative culture of intravenous catheters and other intravascular inserts. J Infect Dis 1980; 141: 781-786.

16.   Brun-Buisson C, Abrouk F, Legrand P, Huet Y, Larabi S, Rapin M. Diagnosis of central venous catheter-related sepsis. Critical level of quantitative tip cultures. Arch Intern Med 1987; 147: 873-877.

17.   Sherertz RJ, Raad II, Belani A, *et al.* Three-year experience with sonicated vascular catheter cultures in a clinical microbiology laboratory. J Clin Microbiol 1990; 28: 76-82.

18.   Farr BM, Shapiro D. Diagnostic tests: Distinguishing good tests from bad and even ugly ones. Infect Control Hosp Epidemiol 2000; 21: 278-284.

19.   Siegman-Igra Y, Anglim AM, Shapiro DE, Adal KA, Strain BA, Farr BM. Diagnosis of vascular catheter-related bloodstream infection: a meta-analysis. J Clin Microbiol 1997; 35: 928-936.

20.   Kite P, Dobbins BM, Wilcox MH, McMahon MJ. Rapid diagnosis of central-venous-catheter-related bloodstream infection without catheter removal. Lancet 1999; 354: 1504-1507.

21.   Sherertz RJ, Heard SO, Raad II. Diagnosis of triple-lumen catheter infection: comparison of roll plate, sonication and flushing methodologies. J Clin Microbiol 1997; 35: 641-646.

22.   Valles J, Rello J, Matas L, *et al.* Impact of using an indwelling introducer on diagnosis of Swan-Ganz pulmonary artery catheter colonization. Eur J Clin Microbiol Infect Dis 1996; 15: 71-75.

23.   Blot F, Chachaty E, Raynard B, Antoun S, Bourgain JL, Nitenberg G. Mechanisms and risk factors of infection of pulmonary artery catheters and introducer sheaths in cancer patients admitted to an intensive care unit. J Hosp Infection 2001; 48: 289-297.

24.   Bjornson HS, Colley R, Bower RH, Duty VP, Schwartz-Fulton JT, Fisher JE. Association between microorganism growth at the catheter insertion site and colonization of the catheter in patients receiving total parenteral nutrition. Surgery 1982; 92: 720-725.

25.   Guidet B, Nicola I, Barakett V, *et al.* Skin versus hub cultures to predict colonization and infection of central venous catheter in intensive care patients. Infection 1994; 22: 43-52.

26.   Mahé I, Fourrier F, Roussel-Delvallez M, Martin G, Chopin C. Predictive value of skin culture in central venous catheter colonization. Réan Urg 1998; 7: 17-24.

27.   Raad II, Baba M, Bodey GP. Diagnosis of catheter-related infections: the role of surveillance and targeted quantitative skin cultures. Clin Infect Dis 1995; 20: 593-597.

28.   Fortun J, Perez-Molina JA, Asensio A, Calderon C, Casado JL, Mir N, Moreno A, Guerrero A. Semiquantitative culture of subcutaneous segment for conservative

diagnosis of intravascular catheter-related infection. J Parenter Enteral Nutr 2000; 24: 210-214.

29. Sitges-Serra A, Puig P, Linares J, Perez JL, Farrero N, Jaurrieta E, Garau J. Hub colonization as the initial step in an outbreak of catheter-related sepsis due to coagulase negative staphylococci during parenteral nutrition. J Parenter Enteral Nutr 1984; 8: 668-672.

30. Segura M, Llado L, Guirao X, Piracés M, Herms R, Alia C, Sitges-Serra A. A prospective study of a new protocol for in situ diagnosis of central venous catheter related bacteraemia. Clin Nutr 1993; 12: 103-107.

31. Fan ST, Teoh-Tchan CH, Lau KF, Chu KW, Kwan AKW, Wong KK. Predictive value of surveillance skin and hub cultures in central venous catheter sepsis. J Hosp Infect 1988; 12: 191-198.

32. Flynn PM, Shenep JL, Strokes D, Barrett FF. Differential quantitation with a commercial blood culture tube for diagnosis of catheter-related infection. J Clin Microbiol 1988; 26: 1045-1046.

33. Andremont A, Paulet R, Nitenberg G, Hill C. Value of semiquantitative cultures of blood drawn through catheter hubs for estimating the risk of catheter tip colonization in cancer patients. J Clin Microbiol 1988; 26: 2297-2299.

34. Douard MC, Clementi E, Arlet G, et al. Negative catheter tip culture and diagnosis of catheter-related bacteremia. Nutrition 1994; 10: 397-404.

35. Capdevila JA, Planes AM, Palomar M, et al. Value of differential quantitative blood cultures in the diagnosis of catheter-related sepsis. Eur J Clin Microbiol Infect Dis 1992; 11: 403-407.

36. Flynn PM, Shenep JL, Strokes DC, Barrett FF. "In situ" management of confirmed central venous catheter-related bacteremia. Pediatr Infect Dis 1987; 6: 729-734.

37. Raucher HS, Hyatt AC, Barzilai A, Harris MB, Weiner MA, LeLeiko NS, Hodes DS. Quantitative blood cultures in the evaluation of septicemia in children with Broviac catheters. J Pediatr 1984; 104: 29-33.

38. Quilici N, Audibert G, Conroy MC, et al. Differential quantitative blood cultures in the diagnosis of catheter-related sepsis in intensive care units. Clin Infect Dis 1997; 25: 1066-1070.

39. Blot F, Nitenberg G, Chachaty E, et al. Diagnosis of catheter-related bacteraemia: a prospective comparison of the time to positivity of central vs. peripheral blood cultures. Lancet 1999; 354: 1071-1077.

40. Blot F, Schmidt E, Nitenberg G, et al. Earlier positivity of central-venous versus peripheral-blood cultures is highly predictive of catheter-related sepsis. J Clin Microbiol 1998; 36: 105-109.

41. Malgrange VB, Escande MC, Theobald S. Validity of earlier positivity of central venous blood cultures in comparison with peripheral blood cultures for diagnosing catheter-related bacteremia in cancer patients. J Clin Microbiol 2001; 39: 274-278.

42. Rijnders BJA, Verwaest C, Peetermans WE, et al: Difference in time to positivity of hub-blood versus nonhub-blood cultures is not helpful for the diagnosis of catheter-related bloodstream infection in critically ill patients. Crit Care Med 2001; 29: 1399-1403.

43. Blot F. Why should paired blood cultures not be useful for diagnosing catheter-related bacteremia in critically ill patients ?. Crit Care Med 2002; 30: 1402-1403.
44. Seifert H, Cornely O, Seggewiss K, *et al.* Bloodstream infection in neutropenic cancer patients related to short-term nontunnelled catheters determined by quantitative blood cultures, differential time to positivity, and molecular epidemiological typing with pulsed-field gel electrophoresis. J Clin Microbiol 2003; 41: 118-123.
45. Tighe MJ, Kite P, Thomas D, Fawley WN, McMahon MJ. Rapid diagnosis of catheter related sepsis using the acridine-orange leukocyte cytospin test and an endoluminal brush. JPEN 1996; 20: 215-218.
46. Rushforth JA, Hoy CM, Kite P, Puntis JW. Rapid diagnosis of central venous catheter sepsis. Lancet 1993; 342: 402-403.
47. Moonens F, El Alami S, van Gossum A, Struelens MJ, Serruys E. Usefulness of Gram staining of blood collected from total parenteral nutrition catheter for rapid diagnosis of catheter-related sepsis. J Clin Microbiol 1994; 32: 1578-1579.
48. Cook D, Randolph A, Kernerman P, Cupido C, King D, Soukup C, *et al.* Central venous catheter replacement strategies: a systematic review of the literature. Crit Care Med 1997; 25: 1417-24.
49. Centers for Disease Control and Prevention. Guidelines for the prevention of intravascular catheter-related infections. MMWR 2002; 51: 1-29.
50. Farr BM. Accuracy and cost-effectiveness of new tests for diagnosis of catheter-related bloodstream infections. Lancet 1999; 354: 1487-1488.

Chapter 5

THE IMPACT OF CATHETER-RELATED INFECTION IN THE CRITICALLY ILL

Christian Brun-Buisson, M.D.
Department of Intensive Care and Infection Control Unit, Centre Hospitalier Universitaire Henri Mondor, Assistance Publique Hôpitaux do Paris and Université Paris, Paris, France

Introduction

Catheter-related infections (CRI) are the third most frequent source of intensive care unit-acquired infection, after pneumonia and urinary tract infection. Moreover, these preventable infections are the most frequent cause of nosocomial bloodstream infection (NBSI). In a recent review, Mermel estimated that about 80,000 cases of bacteremia occurred annually in US intensive care units (ICUs), and might be responsible for 5,000 to 10,000 deaths (1). Catheter-related bacteremia (CRB) also incur morbidity and costs associated with increased length of stay and the treatment of such infections and their complications (2). However, the precise impact of these infections is difficult to delineate and the literature is conflicting, especially with regards to mortality. Difficulties arise with case definitions and methods of analysis.

MORTALITY ASSOCIATED WITH CRI

A number of epidemiological studies document a lower mortality associated with primary and catheter-related bacteremia as compared to secondary bacteremia (3-5). In the study by Pittet and Wenzel, including 364 hospitalized patients with NBSI, secondary bacteremia were associated with an adjusted odds ratio of mortality of 2.46 (95% CI, 1.50-4.02), after accounting for age, underlying diseases (cancer and chronic cardiovascular disease), and polymicrobial infection (4).   In a larger study of 1745 hospitalized patients with NBSI recorded between 1986 and 1991, this same group of investigators found that the crude hospital mortality associated with primary bacteremia and CRB was 30% (and only 25% in definite CRB), whereas it was 43% for secondary bacteremia (5). Mortality at 28 days was associated with age (OR=1.02 per year), cancer or digestive tract disease (OR=1.60),   length of stay before occurrence of bacteremia (OR=1.01 per day), pneumonia as a source (OR=2.74), as well as polymicrobial infection (OR=1.68). In addition, infection with coagulase-negative staphylococci (CoNS) (the major current microbial etiology of CRB) was associated with a favorable outcome (OR=0.79; 95%CI 0.59-1.06), and infection with *Pseudomonas aeruginosa* tended to be associated with an unfavorable outcome (OR=1.04; 95% CI 0.96-1.13). In our own multicenter study, restricted to ICUs and including 111 episodes of NBSI recorded over a four-month period, secondary bacteremia was associated with a similarly higher risk of death (OR = 2.2; 95% CI 0.92-5.15) as compared to primary and catheter-related bacteremia after adjustment for age, severity of underlying disease, and the admission severity score SAPSII; the crude mortality of NBSI was 54% (6). After segregating patients with secondary, primary, and definite catheter-related bacteremia, we found their attributable mortality to be 54.8%, 28.6%, and 11.5%, respectively. Interestingly, Gatell *et al.* (3) found that infection due to "high-risk" sources, including intra-abdominal, lower respiratory tract and primary,- non-catheter-related, bacteremia was independently associated with mortality.

Clearly, one of the problems to evaluate the true impact of catheter-related bacteremia in cohort studies or in case series such as many of those cited above, is that CRB is often mixed with "primary" bacteremias. While this may be appropriate in non-critically ill patients, it may not be in ICU patients, where "true primary" (non-catheter-related) bacteremia in not uncommon. Several of the studies cited above actually suggest that those (non-catheter-related) primary bacteremia have a different outcome from definite CRB. Taken together, these studies would strongly suggest that primary bacteremia is associated with an approximately twice lower risk of death as compared to secondary bacteremia and, also, that among these "primary" bacteremia, CRB is associated with a lower risk of death than (true) "primary" bacteremia, probably also twice or more lower. There are several potential explanations for this. As opposed to other sources (and true primary bacteremia), CRB is associated with a source that can be easily controlled most of the time, the microbial etiology is usually readily known and does not pose therapeutic management dilemma with currently available therapy, and a majority are associated with low-virulence organisms such as CoNS.

A number of studies have focused on catheter-related infection, using various definitions. In 1993, Arnow *et al*, reviewed a case series of catheter-related infections to examine the consequences of catheter-related infections (7). They reported a crude mortality of only 1% in their own series of 102 cases, and an incidence of 20% of complications (i.e., septic metastases). Interestingly, the rate of complications in the papers reviewed was fairly consistent (ranging from 20 to 32%), whereas the mortality rate was not (0% to 23%, with a median value of about 10%). In a study evaluating the impact of CRB in a cohort of surgical ICU patients, 9 of 27 patients with CRB died within 72 h of infection (8). Studies examining attributable mortality must use a case-control design. However, there are clearly problems with evaluating mortality attributable to any specific nosocomial infection occurring in critically ill patients because of competing risks factors for mortality, associated with underlying diseases, acute illness, and other complications of intensive care, including infections at other sites. As a striking parallel (not unrelated) to CRB, the mortality attributable to nosocomial MRSA bacteremia (about 50% of which are CRB) in two recent case-control studies has also varied from 0% (9) to 23% (10). However, there were important differences between the details of the methodology used in the two studies. Harbarth *et al*. studied hospital-wide bacteremia, whereas Blot *et al*. studied ICU-acquired bacteremia. Harbarth *et al*. used a number of detailed and weighted criteria for matching cases and controls (including age, severity of underlying disease, comorbidities, and length of stay prior to bacteremia); a matching score was used to assess the comparability of the two groups, and 12.7 point concordance was achieved by controls out of a

16-point maximum score.8 Blot *et al.* compared the results of two matched cohorts, one comparing MSSA bacteremia to no bacteremia, and another comparing MRSA bacteremia to no bacteremia. In these two series, adjustment was performed only on severity of acute illness (as assessed by the APACHE II score) and diagnostic category, but not on underlying diseases and length of stay prior to bacteremia (10). One possible explanation for the markedly discrepant results between the two studies may be that Harbarth *et al.* have "overmatched" their cases and controls, while Blot *et al* may have "undermatched" their patients.

Similar difficulties arise when attempting to estimate the mortality specifically attributable to CRB. A reference study in the area of NBSI is the cohort study performed by Pittet *et al.* who matched 86 NBSI to 86 controls, and found a 35% mortality attributable to NBSI (11). However, these authors did not differentiate between catheter-related and non-catheter-related bacteremia. The subset of patients with catheter-related NBSI was described in a subgroup analysis (12). The attributable mortality from the infection in this subpopulation was 25% (nine cases versus four controls died). When only matched case-control pairs who survived bloodstream infection (N=11) were considered among patients with infections of intravenous line origin, cases stayed an additional 6.5 days in the surgical intensive care unit (median stay, 15.5 days for cases versus 9 days for controls); extra costs attributable to the infection averaged US$ 28,690 per survivor. A few recent studies, using a case-control design, have focused on the mortality and length of stay attributable to CRB in critically ill patients (5,13-15). Although their results differ somewhat, they do provide a reasonable estimate of the consequences of those infections. A comparison of these four studies is shown in the table. The mortality attributable to CRB ranged from 0% (13,14) to 25% (unadjusted). In the study by Rello *et al.*, the hospital mortality of cases (22.4%) was even lower, although not significantly so, than that of controls (34.7%, P=0.35) (Table).

An interesting, novel approach has been taken in two of these studies by matching (or adjusting) cases and controls on the severity of illness on the days before bacteremia (14,15). Both of these studies found no significant, attributable mortality to bacteremia after adjustment. The study by Soufir *et al.* is particularly interesting in this regard (15). Using a conventional approach by matching cases and controls on variables recorded on admission, these authors found a relative risk of about 2.0 when comparing patients with CRB to patients without CRB (Table). However, when adjusting for severity of illness on the days before bacteremia (day -3 or day -7), these authors found that the relative risk of mortality was reduced from 2.0 to 1.4 or 1.3. While this approach is appealing, it is questionable, because it carries the risk of "overmatching". It should be recalled that in the studies referenced above, CRB occurred a median of >10 days after admission. During that time, a number of events may have occurred which may modify the risk of death in critically ill patients. In fact, adjusting for severity on the days before bacteremia may result in comparing attributable mortality of CRB to that of pneumonia or acute gastrointestinal bleeding, if all other events are not controlled for.

It seems therefore difficult to conclude on a precise value for attributable mortality of CRB. However, taking all results together (Table), it seems reasonable to conclude that CRB carries a relative risk for mortality of about 1.5 as compared to patients without CRB. It does not appear reasonable to suggest that, for example, *Staphylococcus aureus* bacteremia, even when associated with a removable focus of infection has no attributable mortality, as there are clearly documented consequences of such infections (16).

One factor that clearly influences mortality is the microbial etiology. There is no doubt that CoNS infection is associated with a lower risk than other etiologies, particularly *S. aureus*, Pseudomonas or Candida infection. The relative proportion of cases caused by CoNS in a given series will likely influence the analysis of mortality. It is therefore not very surprising to see that Rello *et al.* found a lower attributable mortality than others (Table) since, in their study, the proportion of CoNS was much higher (65%) than in other studies, and the crude mortality of case patients with CRB due to CoNS was only 19.3%, as compared to 27.8% for other microbial etiologies (14). The relatively high proportion of CoNS infection in the study by DiGiovine *et al.* (13) likely contributed also to their negative findings. Another problem, on which there is little information in the above studies, is the timeliness and adequacy of antimicrobial therapy which clearly may impact on the attributable mortality of any infection.

MORBIDITY AND COSTS

Notwithstanding the disputes on mortality, all investigators agree that CRI incur increased morbidity, length of ICU and hospital stay and therefore costs. Pittet *et al.* found that NBSI was associated with an increased hospital length of stay of 14 days overall, and 24 days in survivors only (including 8 days in the ICU); extra costs were estimated at US$ 40,000 (11) In the subgroup of patients with intravenous catheter-associated BSI it averaged US$ 28,690 per survivor (12). Some of the studies listed in the table have evaluated attributable length of stay and costs associated with CRB, and their results are consistent with those estimates, although giving somewhat lower figures. In a recent study, Dimick *et al.* (16) have estimated length of stay and costs associated with CRB after adjusting for confounding variables (demographics and severity of illness). They found a mean increase in total hospital costs (in 1998 dollars) of US$ 56,167 (95% CI, $ 11,523-$165,735), and increased ICU costs of US$ 71,443 (95% CI, $11,960-$ 195,628), as well as a 22-day increase in hospital length of stay, and a 20-day increase in ICU length of stay. However, patients were not adjusted for duration of stay before bacteremia. An interesting feature of our study in this regard is that we have matched patients on variables associated with a longer than median ICU stay, thus eliminating in part confounding factors associated with long ICU stay patients (6). Our estimates of extra length of stay attributable to CRB (the major component of cost) is somewhat lower than that found by Rello and Dimick for example, and may be closer to the average case patient for that reason.

Table. Four Recent Case Control Studies of Catheter-Related Bacteremia in ICUs.

Study (reference)	DiGiovine(13)	Soufir(15)	Rello(14)	Renaud(6)
Study period	3 yrs (1994-96)	6 yrs (1990-95)	7 yrs (1992-98)	4 months (1998)
Setting	1 ICU	2 ICUs	1 ICU	15 ICUs
No. cases/controls	68/68	38/76	49/49	26/26
Etiologic organisms, %				
CoNS	45.5%	5%	63%	24%
S. aureus	10.5%	51%	10%	14%
Enterococci	12%	0	2%	0
Pseudomonas	NR	19.5%	10%	10%
Candida	7.5%	2.5%	2%	7%
Others	25%	27%	12%	45%
Definition of controls	"no evidence of primary bacteremia during ICU stay"	"no evidence of CRI or bacteremia during ICU stay"	"No CRI during ICU stay"	"No bacteremia during ICU stay"
Matching criteria:				
Age	Yes	Optional	Yes	Yes
Severity score ICU admission	No	SAPS II	Yes (APACHE II)	Yes
Severity score < bacteremia	predicted mortality (d-1)	*	No	No
Duration of stay < bacteremia	Yes	Yes	Yes	Yes
Comorbidities	Yes	McCabe class	Yes (chronic disease)	Yes
Admission diagnosis	Yes	Yes	Yes	Yes
Other variables	Date admission	No*	No	No
Crude mortality cases/controls	ICU. 35.3% / 30.9%	ICU .50% / 21.5%	ICU 18.5% / 28.5%	38.5% / 26.9%
Overall attributable mortality	4.4%	28.5%	- 10%	11.5%
Odds ratios	1.33 [0.56-3.16]	2.0 [1.1-3.7]	NR	1.4 [0.86-1.65]
LOS ICU cases/controls, d	13.2 / 5.7	NR		
LOS hospital cases/controls, d	24.2/20.3	NR	63 / 43 (survivors)	32 / 17.5
Attributable LOS hospital, d	4.5	NR	19 ± 49 (survivors)	9.5
Extra costs per episode	US$ 16.000	NR	Euros 3.124	NR

* In this study, matching was performed first as described in the table, then adjustments were made for severity on day-3 (SAPS II, organ dysfunction scores, resulting in lowering relative risk for death in cases as compared to controls (see text).** ICU = intensive care unit

+ LOS = length of stay

++ d = days

To summarize, I have no doubt that there is some excess mortality attributable to the most severe form of catheter-related infection, i.e., CRB; the relative risk can be estimated at about 1.5, or 10% excess mortality on average. This impact may be lower with a higher proportion of infection caused by CoNS as opposed to other organisms and with early appropriate therapy, when needed. Fortunately, CoNS are currently the major etiology of CRB, and the incidence of catheter-related infection appears to have declined in the past decades with improved care and prevention (18). However, the impact of non-bacteremic CRI has not been studied adequately. There is also no doubt that these infections incur excess length of ICU and hospital stay which can be estimated in survivors to approximate 5-10 days and 10-20 days, respectively. It is clear that further gains could be made by the further reduction of these potentially preventable infections.

REFERENCES

1.   Mermel LA. Prevention of intravascular catheter-related infections. Ann Intern Med 2000; 132:391-402.
2.   O'Grady NP, Alexander M, Dellinger EP, Gerberding JL, Heard SO, Maki DG *et al.* Guidelines for the prevention of intravascular catheter-related infections. Centers for Disease Control and Prevention. MMWR Recomm Rep 2002; 51(RR-10):1-29.
3.   Gatell JM, Trilla A, Latorre X, Almela M, Mensa J, Moreno A *et al.* Nosocomial bacteremia in a large Spanish teaching hospital: analysis of factors influencing prognosis. Rev Infect Dis 1988; 10:203-210.
4.   Pittet D, Li N, Wenzel RP. Association of secondary and polymicrobial nosocomial infection with higher mortality. Eur J Clin Microbiol Infect Dis 1993; 12:813-819.
5.   Pittet D, Li N, Woolson RF, Wenzel R.P. Microbiological factors influencing the outcome of nosocomial blodstream infections: a 6-year validated, population-based model. Clin Infect Dis 1997; 24:1068-1078.
6.   Renaud B, Brun-Buisson C, the ICU-Bacteremia Study Group. Outcomes of primary and catheter-related bacteremia. A cohort and case-control study in critically ill patients. Am J Respir Crit Care Med 2001; 163:1584-1590.
7.   Arnow C, Quimosing EM, Beach M. Consequences of intravascular catheter sepsis. Clin Infect Dis 1993; 16:778-784.

8. Charalambous C, Swoboda SM, Dick J, Perl T, Lipsett PA. Risk factors and clinical impact of central line infections in the surgical intensive care unit. Arch Surg 1998; 133:1241-1246.

9. Harbarth S, Rutschmann O, Sudre P, Pittet D. Impact of methicillin resistance on the outcome of patients with bacteremia caused by Staphylococcus aureus. Arch Intern Med 1998; 158:182-189.

10. Blot SI, Vandewoude KH, Hoste EA, Colardyn FA. Outcome and attributable Mmortality in critically ill patients with bacteremia involving methicillin-susceptible and methicillin- resistant Staphylococcus aureus. Arch Intern Med 2002; 162:2229-2235.

11. Pittet D, Tarara D, Wenzel RP. Nosocomial bloodstream infection in critically ill patients. Excess length of stay, extra costs, and attributable mortality. JAMA 1994; 271:1598-1601.

12. Pittet D, Wenzel RP. Nosocomial bloodstream infections in the critically ill. (Letter). JAMA 1994; 272:1820.

13. DiGiovine B, Chenoweth C, Watts C, Higgins M. The attributable mortality and costs of primary nosocomial bloodstream infections in the intensive care unit. Am J Respir Crit Care Med 1999; 160:976-981.

14. Rello J, Ochagavia A, Sabanes E, Roque M, Mariscal D, Reynaga E *et al.* Evaluation of outcome of intravenous catheter-related infections in critically ill patients. Am J Respir Crit Care Med 2000; 162:1027-1030.

15. Soufir L, Timsit JF, Mahe C, Carlet J, Regnier B, Chevret S. Attributable morbidity and mortality of catheter-related septicemia in critically ill patients: a matched, risk-adjusted, cohort study [see comments]. Infect Control Hosp Epidemiol 1999; 20:396-401.

16. Fowler VG, Sanders LL, Kong LK, McClelland RS, Gottlieb GS, Li J *et al.* Infective endocarditis due to Staphylococcus aureus: 59 prospectively identified cases with follow-up. Clin Infect Dis 1999; 28:106-114.

17. Dimick JB, Pelz RK, Consunji R, Swoboda SM, Hendrix CW, Lipsett PA. Increased resource use associated with catheter-related bloodstream infection in the surgical intensive care unit. Arch Surg 2001; 136:229-234.

18. CDC. Nosocomial Infections Surveillance Activity, Hospital Infection Program, National Center for Infectious Diseases. Monitoring hospital-acquired infections to promote patient safety - United States, 1990-1999. MMWR 2000; 49: 149-153. (Erratum p.190).

Chapter 6

THE IMPACT OF CATHETER-RELATED BLOODSTREAM INFECTIONS

Karin E. Byers, M.D., M.S., Barry M. Farr, M.D., M.Sc.
University of Pittsburgh Medical Center, Pittsburgh, PA and the University of Virginia Health System, Charlottsville, Virginia

Case Report

A 44 year-old man without significant past medical illness was thrown from a horse during a fly-fishing trip causing several broken ribs. He attempted to "tough it out" without medical care for 3 days, but his pain became so intense that he was airlifted to a nearby hospital. In the process a peripheral intravenous catheter (IV) was inserted into a vein in his hand while out in the field. This IV was left in place upon arrival to the hospital where he was found to have several broken ribs on the left and a lacerated spleen. He did well initially but then 3 days after admission developed fever to 102.6° F. The next day he was noted to have fever to 104° F and phlebitis at the original hand IV site (now 4 days old). The IV was removed and another IV placed in another location. No antibiotics were started. The next day his temperature went to 105° F. An infectious diseases consultant saw the patient and was able to strip pus from the original IV site. Antibiotic therapy was started. Cultures of blood and of pus from the IV site taken at that time both grew *Staphylococcus aureus*. Shortly thereafter his clinical status began deteriorating rapidly with hypotension and adult respiratory distress syndrome. He was transferred to the intensive care unit (ICU) but died the following day. Autopsy was consistent with suppurative thrombophlebitis and *S. aureus* sepsis was believed to have been the cause of death. [Sherertz RJ, personal communication, November 1, 2003]

Introduction

Catheter-related bloodstream infections (CRBSI) are important adverse effects of healthcare because of added morbidity, mortality, and costs. This chapter will focus on the impact of CRBSI as demonstrated in comparative, epidemiologic studies with control groups being similar types of patients who did not develop CRBSI or patients developing nosocomial bloodstream infections secondary to infections in another organ (e.g., lung). Outcomes have been shown to be affected by host factors such as age, comorbidity, severity of underlying illness, immunosuppression, the virulence of infecting pathogens, the type of catheter infected, the parts of the catheter infected, catheter removal, antimicrobial therapy, and adjunctive antibiotic lock therapy. The fact that so many different variables affect the outcome of CRBSI means that outcomes may vary greatly from one group to another and from one study to another.

Issues of Study Design and Statistics

One more variable that has likely affected some of the reported outcomes of CRBSI has been the way in which a particular study was designed or analyzed. The crude case fatality rate of hospital patients with nosocomial bloodstream infection (BSI) over a period of decades has been about 40% (1), but such crude rates include some dying of the BSI and others dying of their underlying illness or of other adverse effects of hospital care such as a pulmonary embolus. Some observational studies have tried to distinguish between deaths clinically or temporally related to the BSI and those occurring after cure of the BSI due to their underlying illness or another iatrogenic complication. Such studies have supplied rates of mortality attributed to BSI. Other epidemiologic studies have attempted to determine the rate of death due to BSI per se by comparing the mortality rate of patients with BSI with the mortality rate of similar patients without BSI in a cohort study, sometimes with a nested case-control analysis. Such comparisons provide attributable mortality rates. It should be noted that such studies can give false negative results due to inadequate sample size and statistical power. A famous article in the *New England Journal of Medicine* a quarter century ago (2) noted that most studies with negative results until that time had failed to mention statistical power and had provided inadequate power to answer the question being addressed.

When studies of this type compare death rates over a two to three month period between the two groups, control selection becomes a critically important issue (i.e., how well matched the two groups really are) since any systematic difference could result in higher death rates unrelated to the BSI. This is an important consideration since most well treated BSIs are over within days not weeks to months; although some infectious complications could potentially result in death months later, such a time interval allows hospital patients with significant underlying illnesses to die of unrelated causes as well.

Studies of attributable mortality can involve matched controls and/or adjustment for other predictors of death, which provide estimates of attributable mortality adjusted for these other predictors. It should be noted that controlled studies can be overmatched, however, and thereby produce a false negative result. Likewise, multivariate models also can be overfitted and provide false negative results because of inadequate sample size and statistical power to address the number of variables included in the model. For the above reasons, two studies of the same question can give different results apparently due to differences in study methods.

These observations about false positive and false negative study results may recall for some Disreali's phrase "lies, damned lies, and statistics." It is clear that one can find a study with inferential statistics and impressive probability values that "proves" pretty much anything one might desire regarding the mortality or complication rates of CRBSI. Austin Bradford Hill proposed that one needs to find consistency of evidence among the preponderance of epidemiologic studies of a question conducted by different investigators in different settings and populations before concluding that a particular association is valid or causal. It also has been suggested that the purpose of biostatistics is to obtain "reliable results" through appropriate data collection, classification and analysis. The objective of this chapter will be to seek the most reasonable answers among sometimes conflicting data regarding the morbidity, mortality and excess costs associated with CRBSI under a variety of different conditions.

Morbidity

The impact felt first and foremost by patients with CRBSI is additional suffering. CRBSI ranges in severity from a relatively mild fever to dramatic, catastrophic illness with high fever (or hypothermia), septic shock, disseminated intravascular coagulation, and adult respiratory distress syndrome (3). It has been estimated that several hundred thousand of the roughly two million, nosocomial infections acquired in U.S. hospitals annually are CRBSIs. Many more healthcare-associated CRBSIs occur now in long-term care facilities and in home care. Because CRBSI is not a reportable illness, these estimates have been extrapolated from many publications of rates from single hospitals and a number of multicenter studies such as those by the National Nosocomial Infection Surveillance (NNIS) program organized and conducted by the Centers for Disease Control and Prevention (CDC).

In addition to the morbidity of the primary illness, some patients suffer a relapse of bloodstream infection after completing a course of therapy. Other complications can include septic thrombophlebitis, endocarditis, or metastatic, hematogenous infections such as epidural abscess, osteomyelitis, or septic arthritis. These complications often require prolonged medical therapy and some also require a surgical operation.

Reported CRBSI complication rates have ranged from 20% to 45%. In the study reporting the highest rate of complications, 71% of the complications were considered to be "major"(3). This included patients with septic shock, sustained sepsis, suppurative thrombophlebitis, metastatic infection, endocarditis and septic arthritis. Infections with *Staphylococcus aureus* were associated with the highest rate of serious complications, which included all episodes of metastatic infections and one episode of endocarditis. In another study that included only patients with Staphylococcus aureus CRBSI, however, only 9 (16%) of 55 developed acute early complications, including endocarditis, fulminant sepsis, osteomyelitis, septic arthritis and septic pulmonary emboli (4).

It is difficult to quantify the amount of additional suffering incurred due to CRBSI and associated complications, but the prolongation of hospital stay perhaps provides an index of this added morbidity since very few patients wish to remain in the hospital when well enough to be discharged.

PROLONGATION OF HOSPITAL STAY

The average increase in hospital stay due to nosocomial bloodstream infection in the CDC's SENIC study conducted during the 1970s and 1980s was 7 days. A more recent study evaluating all types of BSI in a surgical intensive care unit (ICU) reported a similar, median increase of 8 days but the increase was 24 days for patients who survived the BSI .

Several recent studies focusing upon CRBSI reported overall mean increases ranging from 10 to 22 days (4-6). After accounting for severity of disease and demographics in one of these studies, BSI patients still had an associated 124% increase in LOS (22 days, 95% CI 7-70 days, P=.002) (7). In studies of ICU patients, the duration of ICU stay increased by 5 to 20 days for patients with CRBSI. (6,7). A fourth recent study reported no significant increase in stay due to CRBSI, however (8). Differences in study design or analysis may have accounted for this different result, but the bulk of the available studies suggest a substantial prolongation of hospital stay due to CRBSI.

COSTS

In an uncontrolled 1991 study, the average excess cost was estimated to be $3,707 for all episodes of CRBSI and $6,064 for those caused by *Staphylococcus aureus* (3). Another study evaluated the accuracy of such physician estimates by comparing them with the measured costs in patients with CRBSI vs. those of matched controls without CRBSI. The additional charges measured in the comparison were approximately 2.5-fold greater than those estimated by the physician (9).

A later study of CRBSI costs reported that total hospital costs increased by $56,167 in 1998 dollars; for patients in the ICU, additional ICU costs were $71,443 (7). In a multivariate analysis adjusting for severity of illness and demographic factors, CRBSI was still associated with a 120% increase in total hospital costs (95% CI $11,523-$165,735, P=.001). The cost of room and board was the single largest contributor to this increased cost, but costs were increased across all categories except operating room costs.

In a 1999 study, the estimated increase in medical costs for ICU patients with primary bloodstream infections was reported to be more than $16,000 per episode (6).

MORTALITY

Without question, the most important adverse outcome of bloodstream infection (BSI) is death. Nosocomial BSI has been associated with death in many studies over a period of decades (1) and some have suggested that 50,000 U.S. patients die of this each year (10), which would make this a more frequent cause of death than the Acquired Immune Deficiency Syndrome (AIDS) in the United States. This seems to be an overestimate, however, because deaths directly due to CRBSI have been relatively rare in our experience and multiple recent studies have found no association between BSI and mortality at all after adjusting for comorbity and/or underlying severity of illness (11). Some have therefore suggested that patients often die with rather than because of BSI. Adding further intensity to the controversy, primary BSIs due to a removable focus of infection such as a vascular catheter have long been recognized to have lower crude case-fatality rates and have thus been thought to be less deadly than secondary BSIs arising from a serious infection in another organ such as the lung. The case report at the beginning of the chapter convincingly demonstrates death due to vascular catheter-related infection in an otherwise healthy individual. This section will focus on the how often death is due to CRBSI using epidemiologic data regarding the attributable mortality of bloodstream infection in general and CRBSI in particular

Severity of Underlying Illness vs. Severity of the Bloodstream Infection

Patients with higher age, significant comorbidity and severe underlying illnesses are more likely to die of CRBSI (12), but several studies have also shown that severity of the BSI independently predicts death even when these factors are taken into account. (12,13) One of the latter studies showed that the severity of illness as reflected by the APACHE II score early in the BSI was a significant predictor of death (13). Another found that the increase in the patient's APACHE II score early during the BSI was most predictive of death (14). Still another study found that shock was predictive of death and there was a trend toward higher death rates with onset on a hospital ward rather than in an ICU; this did not reach statistical significance in multivariate analysis, perhaps due to the study's small sample size and low statistical power (15).

Primary vs. Secondary Bloodstream Infections

It seems clear from the consistent results of multiple studies that BSIs secondary to serious infections of another organ such as the lung are associated with significantly higher crude mortality than are primary BSIs, most of which are due to a removable focus of infection at the site of a vascular catheter (16,17).

Mortality Due to CRBSI

In 1988, Maki estimated the attributable mortality of CRBSI to be 10-20%. Several studies have reported mortality higher than Maki's estimate ranging from 20%-35% (18). Other studies have found much lower mortality rates from CRBSI, however. Three prospective studies involving 266 patients and 665 central venous catheters at the University of Virginia reported no deaths related to 24 episodes of CRBSI (19-21). Likewise, Capdevilla reported on 40 episodes of CRBSI in hemodialysis patients with no deaths (22). All episodes were treated with systemic antibiotics and antibiotic lock for two weeks; all patients were cured and all hemodialysis catheters were salvaged. In another study of hemodialysis catheter related infections none of 50 patients with *S. aureus* CRBSI died (23). In a study of 102 episodes of catheter-related bacteremia in hospitalized adults, only 1 patient was believed to have died from the infection itself and another 18 were thought to have died primarily of other causes during treatment of IV catheter sepsis. All of those patients had rapidly fatal or ultimately fatal underlying illnesses (3). A meta-analysis of 129 studies representing 2,573 episodes of CRBSI, found a crude mortality of 14% (95%CI 12.4-15.6%), but the authors of the original studies found a clinical, temporal association with death in 2.7% of cases of CRBSI (95% CI 2.0-3.4%) (24). It should be noted that the 129 studies reported "attributed mortality" (i.e., deaths felt to be obviously connected to CRBSI by the clinician investigators who authored the original studies included in the meta-analysis). In several other recent studies, CRBSI was not significantly associated with mortality after adjusting for severity of underlying illness (5-7,25).

Relationship to Therapy

The bulk of the studies reporting on use of antimicrobial therapy inappropriate for the causative agent (i.e., defined as resistance in laboratory susceptibility testing) have found this to be associated with higher mortality even when provided for the first day or so (26). Likewise, antibiotic resistant infections, which are associated with delays in correct therapy and often with use of less active, second line agents have been associated with higher mortality even when adjusted for severity of underlying illness (26-28). Removal of the catheter early after onset of a CRBSI has been associated with higher cure rates and lower relapse rates in a number of studies (29), but others have reported no deaths with appropriate antimicrobial treatment despite a large number of CRBSIs (including many in which the catheter was left in place and salvaged) (19-23). Relapse rates after intraluminal CRBSIs have been lower after parenteral plus antibiotic lock therapy (22).

Relationship to the Type of Catheter Infected

The type of catheter may predict the outcome of infection. This may be due to the patient population requiring a specific type of catheter or to the catheter itself. In a meta-analysis of catheter-related BSI , pulmonary artery catheters were associated with the highest case fatality rate (52.6%, 95% CI 28.9-75.6), followed by peripheral catheters (31.2%, 95% CI 11.0-58.7), totally implanted subcutaneous ports (25%, 95% CI 3.2-6.51), umbilical catheters (15.4%, 95% CI 1.9-45.4), hemodialysis catheters (11%, 95% CI 6.1-15.9), unspecified central venous catheters (9.7%, 95% CI 6.2-13.2), tunneled central venous catheters (9.4%, 95% CI 6.8-12.0) and arterial catheters (7.7%, 95% CI 0.2-36.0) (24). It should be noted that the confidence intervals for some of these rates were very wide.

Relationship to the Etiologic Agent

Some of the organisms associated with catheter infections are considered to be relatively avirulent. These include coagulase-negative staphylococci and diphtheroids. Moreover, several recent studies have reported that most blood cultures growing coagulase negative staphylococci were due to contaminants including quite a few with more than one set of blood cultures positive (i.e., because each set contained a different strain). In a meta-analysis of articles describing 2,573 cases of CRBSI and their outcomes, coagulase-negative staphylococci and enterococci had lower rates of attributed mortality than did other etiologic agents (0.7% and 0%, respectively) (24). By contrast, Candida species and *S. aureus* were each associated with significantly higher attributed mortality rates than were other microbes (9%, p = .001, and 8.2%, p < .001, respectively).

S. aureus is a virulent organism that is often associated with late complications. In a different meta-analysis, the overall complication rate was 24% and the mortality rate attributed to *Staphylococcus aureus* was 15%. Because of a higher risk of late complications than with many other pathogens, the optimal duration of antibiotic therapy for this entity remains unclear (4,30,31). Some have suggested that a 2-week duration of therapy may be adequate for uncomplicated *S. aureus* catheter-related bacteremia while others have maintained that the risk of late complications and relapse makes this approach unacceptable unless a transesophageal echocardiogram is negative and the patient responds quickly to appropriate antimicrobial therapy (4,31). Fever or *S. aureus* bacteremia that persists for more than 3 days after catheter removal and initiation of antibiotic therapy predict early complications, including endocarditis, osteomyelitis, fulminant sepsis, septic arthritis and septic pulmonary emboli. Among patients without early complications, 3 (16%) of 18 treated for less than 10 days had a later relapse. For these reasons, these patients may require a longer course of antibiotics (4,30).

CONCLUSIONS

CRBSIs clearly cause the death of some patients but are less likely to do so than are secondary BSIs. Death is more likely if the patient is already severely ill or receives delayed or incorrect therapy, both of which are significantly more common with antibiotic resistant pathogens. Species like *S. aureus* and Candida species are more likely to cause death. Early severity of the BSI including complications such as septic shock, ARDS and DIC also predict death. There appears to be an emerging consensus that the attributable mortality of CRBSI is probably closer to 3% than to 30%. This is compatible with the results of multiple recent studies finding no significant association with mortality at all, since their confidence intervals included this 3% level and their sample sizes usually did not permit detecting such a small difference from the null hypothesis. Studies reporting high attributable mortality rates have usually not adjusted for patients' comorbidity and underlying severity of illness. Their results are also incompatible with those of a number of other studies of ICU, hemodialysis and bone marrow transplant unit CRBSIs in which there were no deaths. It should be recognized that this low attributable mortality is largely due to the frequency of prompt, appropriate therapy in a large majority of cases of nosocomial CRBSI. This should not be taken to mean that CRBSIs are unimportant. They clearly cause large amounts of morbidity, prolong hospital stay, add significant costs, and sometimes kill the patient.

REFERENCES

1. Ziegler EJ, Fisher CJ Jr, Sprung CL, Straube RC, Sadoff JC, Foulke GE, Wortel CH, Fink MP, Dellinger RP, Teng NN, et al. Treatment of gram-negative bacteremia and septic shock with HA-1A human monoclonal antibody against endotoxin. A randomized, double-blind, placebo-controlled trial. The HA-1A Sepsis Study Group. N Engl J Med 1991; 324(7):429-36.

2. Freiman JA, Chalmers TC, Smith H Jr, Kuebler RR. The importance of beta, the type II error and sample size in the design and interpretation of the randomized control trial. Survey of 71 "negative" trials. N Engl J Med 1978;299:690-94.

3. Arnow PM, Quimosing EM, Beach M. Consequences of intravascular catheter sepsis. Clin Infect Dis 1993;16(6):778-84.

4. Raad II, Sabbagh MF.Optimal duration of therapy for catheter-related *Staphylococcus aureus* bacteremia: a study of 55 cases and review. Clin Infect Dis 1992;14(1):75-82

5. Rello J, Ochagavia A, Sabanes E, Roque M, Mariscal D, Reynaga E, Valles J.Evaluation of outcome of intravenous catheter-related infections in critically ill patients. Am J Respir Crit Care Med 2000 162(3 Pt 1):1027-30.

6. Digiovine B, Chenoweth C, Watts C, Higgins M. The attributable mortality and costs of primary nosocomial bloodstream infections in the intensive care unit. Am J Respir Crit Care Med 1999;160(3):976-81.

7. Dimick JB, Pelz RK, Consunji R, Swoboda SM, Hendrix CW, Lipsett PA.Increased resource use associated with catheter-related bloodstream infection in the surgical intensive care unit. Arch Surg. 2001 Feb;136(2):229-34.

8. Pelletier SJ, Crabtree TD, Gleason TG, Pruett TL, Sawyer RG. Bacteremia associated with central venous catheter infection is not an independent predictor of outcomes.J Am Coll Surg 2000;190(6):671-80; discussion 680-1.

9. Haley RW, Schaberg DR, Von Allmen SD, McGowan JE Jr. Estimating the extra charges and prolongation of hospitalization due to nosocomial infections: a comparison of methods. J Infect Dis 1980;141(2):248-57.

10. Wenzel RP, Edmond MB. The evolving technology of venous access. N Engl J Med 1999 7;340(1):48-50.
11. Blot S, Vandewoude K, Hoste E, Colardyn F. Reappraisal of attributable mortality in critically ill patients with nosocomial bacteraemia involving Pseudomonas aeruginosa. J Hosp Infect 2003;53(1):18-24.
12. Lesens O, Methlin C, Hansmann Y, Remy V, Martinot M, Bergin C, Meyer P, Christmann D.Role of comorbidity in mortality related to *Staphylococcus aureus* bacteremia: a prospective study using the Charlson weighted index of comorbidity. Infect Control Hosp Epidemiol 2003;24(12):890-896.
13. Harbarth S, Ferriere K, Hugonnet S, Ricou B, Suter P, Pittet D.Epidemiology and prognostic determinants of bloodstream infections in surgical intensive care. Arch Surg 2002;137(12):1353-9;
14. Yzerman EP, Boelens HA, Tjhie JH, Kluytmans JA, Mouton JW, Verbrugh HA. Delta APACHE II for predicting course and outcome of nosocomial *Staphylococcus aureus* bacteremia and its relation to host defense. J Infect Dis 1996;173(4):914-9.
15. Lundberg JS, Perl TM, Wiblin T, Costigan MD, Dawson J, Nettleman MD, Wenzel RP. Septic shock: an analysis of outcomes for patients with onset on hospital wards versus intensive care units. Crit Care Med 1998;26(6):1020-4.
16. Pittet D, Li N, Wenzel RP. Association of secondary and polymicrobial nosocomial bloodstream infections with higher mortality. Eur J Clin Microbiol Infect Dis 1993;12(11):813-9.
17. McGregor AR, Collignon PJ. Bacteraemia and fungaemia in an Australian general hospital--associations and outcomes. Med J Aust 1993;158(10):671-4.
18. Collignon PJ. Intravascular catheter associated sepsis: a common problem. The Australian Study on Intravascular Catheter Associated Sepsis. Med J Aust 1994;161(6):374-8.
19. Cobb DK, High KP, Sawyer RG, Sable CA, Adams RB, Lindley DA, Pruett TL, Schwenzer KJ, Farr BM. A controlled trial of scheduled replacement of central venous and pulmonary-artery catheters. N Engl J Med 1992;327(15):1062-8.
20. Flowers RH 3rd, Schwenzer KJ, Kopel RF, Fisch MJ, Tucker SI, Farr BM. Efficacy of an attachable subcutaneous cuff for the prevention of intravascular catheter-related infection. A randomized, controlled trial. JAMA 1989;261(6):878-83.
21. Howell PB, Walters PE, Donowitz GR, Farr BM. Risk factors for infection of adult patients with cancer who have tunnelled central venous catheters. Cancer 1995;75(6):1367-75.
22. Capdevila JA, Segarra A, Planes AM, Gasser I, Gavalda J, Pahissa A. Long term follow-up of patients with catheter related sepsis (CRS) treated without catheter removal [Abstract J3]. In: Programs and Abstracts of the 35th Interscience Conference of Antimicrobial Agents and Chemotherapy. San Francisco, 1995.
23. Peacock SJ, Curtis N, Berendt AR, Bowler IC, Winearls CG, Maxwell P. Outcome following haemodialysis catheter-related *Staphylococcus aureus* bacteraemia. J Hosp Infect 1999;41(3):223-8.
24. Byers KE, Adal KA, Anglim AM, Farr BM. Case fatality rate for catheter-related bloodstream infections (CRBSI): a meta-analysis. Infect Control Hosp Epidemiol 1995; 16(Part 2, Suppl):23.

25.   Soufir L, Timsit JF, Mahe C, Carlet J, Regnier B, Chevret S. Attributable morbidity
       and mortality of catheter-related septicemia in critically ill patients: a matched, risk-
       adjusted, cohort study. Infect Control Hosp Epidemiol 1999;20(6):396-401.
26.   Ibrahim EH, Sherman G, Ward S, Fraser VJ, Kollef MH. The influence of inadequate
       antimicrobial treatment of bloodstream infections on patient outcomes in the ICU
       setting. Chest 2000;118(1):146-55.
27.   Blot SI, Vandewoude KH, Hoste EA, Colardyn FA. Outcome and attributable
       mortality in critically Ill patients with bacteremia involving methicillin-susceptible
       and methicillin-resistant *Staphylococcus aureus*. Arch Intern Med
       2002;162(19):2229-35.
28.   Cosgrove SE, Sakoulas G, Perencevich EN, Schwaber MJ, Karchmer AW, Carmeli
       Y. Comparison of mortality associated with methicillin-resistant and methicillin-
       susceptible *Staphylococcus aureus* bacteremia: a meta-analysis. Clin Infect Dis
       2003;36(1):53-9.
29.   Fowler VG Jr, Sanders LL, Sexton DJ, Kong L, Marr KA, Gopal AK, Gottlieb G,
       McClelland RS, Corey GR. Outcome of *Staphylococcus aureus* bacteremia according
       to compliance with recommendations of infectious diseases specialists: experience
       with 244 patients. Clin Infect Dis 1998;27(3):478-86.
30.   Fowler VG, Olsen M, Corey GR, Woods C, Reller A, Cheng J *et al*. Predictors of
       complications in patients with *Staphylococcus aureus* bacteremia [Abstract 415]. In:
       Abstracts of the 41st Interscience Conference on Antimicrobial Agents and
       Chemotherapy September and December 2001; 415.
31.   Fowler VG Jr, Li J, Corey GR, Boley J, Marr KA, Gopal AK, Kong LK, Gottlieb G,
       Donovan CL, Sexton DJ, Ryan T. Role of echocardiography in evaluation of patients
       with *Staphylococcus aureus* bacteremia: experience in 103 patients. J Am Coll
       Cardiol 1997;30(4):1072-8.

Chapter 7

MANAGEMENT AND TREATMENT

Amar Safdar, M.D. and Issam I. Raad, M.D.
The University of Texas M.D. Anderson Cancer Center, Houston, Texas

Introduction

Catheter-related bloodstream infections (CRBSIs) have evolved as important systemic infections due to a near-universal use of intravacular devices in the critically ill hospitalized patients (1). Hematogenous bacterial and fungal nososcomial infections attributed to an infected indwelling catheter are associated with significant morbidity and near-exponential increase in health-care expenditure (2,3). Most (86%) of nosocomial BSIs are attributed to infected catheter source (4,5), this in most part has led to a substantial rise in coagulase-negative staphylococci bloodstream infection, which has become the leading cause of nosocomial bacteremia in the hospitalized patients (6). Emergence and rise in the multidrug-resistant *Staphylococcus aureus* in hospitalized patients (7) and those from community (8) has also posed serious challenges in selection of appropriate empiric antimicrobial therapy for patients with CRBSI. It is therefore imperative that high-level of suspicion, prompt diagnosis and appropriate therapy be instituted early, especially in severely ill patients with infected intravascular device-related bacteremia or fungemia.

COLONIZATION VERSUS INFECTION

Nearly all, indwelling, intravascular catheters may develop a biofilm layer within 24 h after insertion; these biofilms provide a protective milieu for viable bacterial existence in sessile form (9,10). Microorganisms embedded in these biofilm layers may yield to planktonic form or free-floating bacteria on the catheter surface that increases the risk of hematogenous invasion and systemic dissemination (1). A well established association between high-grade bacterial inoculum at the catheter source (semiquantitative or quantitative cultures) and increased potential of bloodstream invasion (11,12), form the bases of diagnostic criteria for catheter-related infection versus colonization. A five-fold higher yield from catheter blood sample compared to percutaneous blood culture taken at the same time is currently use to defined CRBSI.

Initial bacterial colonization of catheter depends upon multiple factors: (a) microorganism-related cell-surface characteristics for example, hydrophobic organisms like Staphylococcus adhere better to polyvinyl chloride, silicone, and polyethylene surface compared to Teflon polymers or polyurethane (13). (b) Catheter surface such as surface irregularities, difference in electric charges may result in enhance bacterial adhesion, and perhaps selection of microorganisms. (c) Host-derived proteins in the bioflim matrix such as fibronectin prefers *Staphylococcus epidermidis*, whereas *Staphylococcus aureus* uses a wider variety of host proteins like fibronectin, fibrinogen and to a lesser extent laminin to anchor and harbor in the "slime" matrix. (d) Bacterial derived enzymes resulting from phenotypic switch following changes in environment may also contribute in increased biofilm formation (14), which in turn leads to enhanced ability of bacteria to adhere to the catheter surface and promote extracellular "slime" or exopolysaccharide formation. This further conforms to the bifilm embedded microorganism a favorable microenvironment by compromising efficacy of antimicrobial drug, especially of glycopeptide-based antibiotics (15).

ETIOLOGY OF BLOODSTREAM INVASION

Microorganisms associated with CRBSI are generally acquired from the skin flora at the catheter insertion site or through the hub of the catheter device (16). Infections of catheter placed for a short-term (< 10 days) duration often results from bacterial migration along the outer catheter surface from the insertion site (17), whereas chances of bacterial contamination of the catheter hub are increased if catheters are left in place for prolonged periods (> 30 days); in the later case, organisms migrate along the inner surface of the catheter and lead to either colonization of the biofilm matrix and/or hematogenous invasion/infection.

Microorganisms involved in CRBSIs are mostly those associated with skin flora around the catheter insertion site, such as *Staphylococcus epidermidis*, *S. aureus*, *Bacillus* speices, *Cornybacterium* species; in patients with groin lines aerobic gram-negative bacteria and enterococci may be occasionally encountered. Infections due to *Pseudomonas aeruginosa*, *Acinetobacter* species, *Stenotrophomonas maltophilia*, *Candida* species, especially *C. parapsilosis* has been occasionally attributed to iatrageneic catheter contamination (12,18). Other organisms infrequently encountered include, *Micrococcus* speices, viridance Streptococci, especially in patients with severe oropharygeal mucositis; catheter hub infections due to rapidly growing mycobacteria such as *Mycobacterium fortuitum*, *Mycobacterium chelonei* and *Mycobacterium abscessus* are rare. In immunosuppressed individuals CRBSI due to *Fusarium* species, *Rhodotorula* species, *Hansenula anomala*, *Trichosporon beigelii* and other *Trichosporon* species may cause recalcitrant catheter-related fungemia (17–21). *Malassezia furfur* infections are often associated with lipid component of hyperalimentation and in patients with severe mucocutaneous excoriation (22).

MICROORGANISM-DIRECTED THERAPY

The treatment modality and anticipated response to therapy for infections associated with indwelling intravascular devices may vary significantly based on the microorganism involved (23). In this section pathogen-directed approaches in the management of CRBSIs are discussed.

Coagulase Negative Staphylococcus (CNS)

Catheter-related infections has significantly increased the incidence of nosocomial bacteremia due to *S. epidermidis*, which has become the most common cause of CRBSI (24). Most patients are febrile on presentation albeit, sepsis like clinical picture is often absent in this setting. Presence of fever, blood culture > 1, preferably from two separate sites and absence of other sources of fever indicates that contamination is less likely source of CNS blood culture isolation.

Antimicrobial Therapy

- In non-persistent infection, antimicrobial therapy alone is often adequate and catheter removal is not routinely warranted, especially in individuals with long-term tunneled catheters. Albeit, catheter retention has been associated with high rate of infection relapse (20%) compared to patients that had CVC removed (25). In the event CVC is retained, duration of antibiotic therapy is extended from 5–7 days to 10–14 days.
- In patients with complicated CNS CRBSI, all intravascular lines must be removed promptly, and duration of systemic antimicrobials for endovascular infection is 4–6 weeks; acute CNS osteomyelitis treatment is given for 6–8 week.

Staphylococcus aureus

Catheter-related *S. aureus* bacteremia poses a serious challenge, as patients with *S. aureus* bloodstream infections are at a higher risk of endovascular infection such as endocarditis, and septic thrombophelibitis, as well as distant suppurative complications, such as osteomyelitis, and deep tissue abscesses (26,27). Fever may be the initial presentation although clinical signs of sepsis are not infrequent in this setting. Optimum duration of therapy is not known, and prolonged (> 14 days) antimicrobial therapy is often recommended due to the probability of unacceptably higher rate of secondary suppurative complications. Higher rate of infection relapse has been observed in patients treated with short courses of antibiotic (23). Several recent studies including author's experience (28), indicates that appropriate antimicrobials when given for a shorter duration (10–14 days) may be an acceptable alternative in patients with clinically uncomplicated catheter-related *S. aureus* bacteremia; an observation further supported by meta-analysis of 11 studies that showed 6% relapsed in patients that received short-course antibiotic therapy (29).

Antimicrobial Therapy

- We recommend that patients with *S. aureus* CRBSI that shows prompt clinical response to therapy may be eligible to receive a shorter duration (~ 14 days) of parentral antibiotic therapy, bactericidal b-lactams are preferred and in non-b-lactam susceptible isolates, vancomycin is often the agent of choice although the newer agents (quinupristin/dalfopristin; linezolid) appear promising and need further evaluation. Clinical response in patients with CRBSI is defined as (a) abatement of fever (< 3 days); (b) sterile blood cultures within 72 h after antibiotic therapy is commenced; (c) No evidence of endocarditis using modified Duke diagnositic criteria (30).
- Transesophageal echocardiogram (TEE). Patients with a known heart valve disorder and those with prosthetic endovascular device, such as pacemaker, artificial heart valve, and foreign vascular grafts may require preemptive TEE. In others, TEE may be performed according to treating physician's discretion, especially in individuals that remain febrile and/or bacteremic for > 48 hrs despite appropriate antibiotic therapy. Due to low sensitivity, high coast, and potential morbidity, TTE is not routinely recommended for all patients with *S. aureus* CRBSI.
- CVC Removal. Catheter removal has been associated with rapid response and reduced rate of infection recurrence (31). This approach is often impractical in patients with difficult to access surgically implanted, tunneled indwelling catheters. Currently, hybrid protocols address removal of an infected catheter in patients with *S. aureus* CRBSI, retention of infected CVC has been accomplished with prolonged systemic antibiotics plus antibiotic lock therapy (discussed later). Antibiotic lock technique in addition to parentral antibiotic therapy may improve cure rate and has shown to salvage infected catheters (32). We recommend that catheter should be removed and antibiotic lock therapy differed if any of the following exists: (a) Multidrug-resistant organism, such as methacillin-resistant *S. aureus* (MRSA); (b) Patients either intolerant or allergic to optimum bactericidal antimicrobial agent; (c) Individuals with increased risk of endovascular infection, which presents as prolonged,

recurrent high-grade bacteremia; (d) Lack of early (< 3 days) clinical and microbiologic resolution of infection after institution of appropriate antimicrobial therapy.

- Catheter salvage. If catheters cannot be removed without serious untoward risk for the patient, extended salvage therapy may be entertained including, minocycline, TMP-SMX, fluoroquinolone plus a once daily dose of rifampin following conventional antimicrobial therapy (23,33,34). If patients are treated with salvage approach than high-level of suspicion for secondary complications including endocarditis, osteomyelitis or distant suppurative processes should be anticipated, diagnosed and treated promptly.

Aerobic Gram-Positive Bacilli

Infections due to other normal skin microflora organisms such as *Corynebacterium jeikeium* and *Bacillus* species are infrequent cause of CRBSI, and treatment approach is quite similar to that of S. epidermidis CRBSI. Vancomycin is the treatment of choice, and short-term non-surgically placed catheter may be removed; in surgically placed intravascular accesses, salvage treatment trial may be attempted in individuals with non-persistent BSI and those with no evidence of exit-site or tunnel infection. Presence of refractory bacteremia (> 72 h) after institution of appropriate antibiotics warrants catheter removal.

Aerobic Gram-Negative Bacilli

Catheter infections due to enteric gram-negative rods such as *Escherichia coli, Klebsiella pneumoniae* are rare and often seen in patients with long-standing groin intravascular catheters. In the immunosuppressed patients receiving antineoplastic chemotherapy, non-*aeruginosa Pseudomonas* species, and *Stenotrophomonas maltophilia* bacteremia are often associated with indwelling catheter infection, and successful treatment includes appropriate systemic antimicrobial therapy along with removal of the infected intravascular device (35). CRBSI due to *Pseudomonas* species, *Acinetobacter* species, *Stenotrophomonas maltophilia* or *Enterobacter agglomerans* group may be associated with either contaminated infusate or catheter hub contamination due to repeat manipulation by medical personal.

Antimicrobial Therapy

- Optimum treatment includes early institution of appropriate antimicrobial agents given for 10 to 14 days, and removal of the indwelling catheter. Patients with to *Pseudomonas* species, *Acinetobacter* species, and *Enterobacter agglomerans* group infection, we recommend combination antibiotic therapy with in vitro synergistic antimicrobial activity, such as b-lactam plus aminoglycoside, or fluoroquinolones. Treatment with high-dose TMP-SMX, and/or ticarcillin-clavulanate may be approached cautiously in catheter-related hematogenous *Stenotrophomonas maltophilia* infections, as organism is frequently resistant to most broad-spectrum agents and often develop de novo resistance, especially if source of infection (CVC) has not been eliminated.
- In catheter salvage approach, extended course of antibiotics may be given along with antibiotic lock therapy, albeit high treatment failure rates and increased recurrence may not be unexpected.

Candida species

Candida species fungemia may be associated with infected intravascular catheter, and frequently seen in oncologic patients undergoing cancer therapy (36). Treatment often involves catheter removal along with systemic antifungal therapy for ~ 14 days following last positive blood culture to prevent infrequent but profoundly disabling complication of permanent vision loss due to yeast endophthalmitis (23,37).

Antimicrobial Therapy

- For *C. albicans*, *C. tropicalis* and *C. parapsilosis* treatment with parental fluconazole or caspofungin may be used as first line therapy (38,39).
- For *C. glabarata* and *C. krusei* CRBSI, caspofungin or amphotericin B are suggested. Clinical efficacy of caspofungin in the non-neutropenic patients with *Candida* species CRBSI

appears promising (39) although further information is needed in recommending changes in treatment guidelines; in profound granulocytopenic patients with catheter-related fungemia antifungal treatment includes amphotericin B and its lipid-based preparations, caspofungin may be used in low-risk patients (neutropenia < 7 days) with caution.

Filamentous Fungi and Other Yeasts

Similar to candidemia CRBSIs due to *Fusarium* species, *Trichosporon* species *Malassezia furfur* requires removal of the infected catheter if other sources of bloodstream infection have been excluded. Treatment often requires systemic amphotericin B, however, for infections due to non-polyene (amphotericin B)-susceptible fungi such as *Fusarium* species optimum antifungal therapy is not clear; we currently use a combination of high-dose AmBisome® plus caspofungin (40).

Rapidly Growing Mycobacteria

Mycobacteremia due to RGM is rare and treatment is based on prompt removal of infected catheter along with 4-week antimicrobial therapy with susceptible drug combination therapy usually consisting of fluoroquinolone plus azolide and rifampin. *Mycobacterium abscesses* tends to have high-level of drug-resistance and routine susceptibility analysis will yield much-needed in vitro guideline for optimum drug combinations.

CATHETER SALVAGE STATAGIES

Long-term surgically placed catheters are not easy to replace and to reduce severe morbidity and health-care expenditure, catheter salvage is attempted in most patients that are stable and do not exhibit signs of clinical decompensation or sepsis with organ-dysfunction, or refractory hypotension. In certain settings attempts to salvage catheter may have low yield, such as tunnel or pocket infection, or infection due to *S. aureus*, *P. aeruginosa*, and *Candida* species (23).

Antibiotic Lock Therapy

The failure of effective antimicrobial elimination of sessile microorganisms that resides within the semi-protective biofilms has led to the development of "antibiotic lock" technique, which requires filling of catheter hub and lumen with higher antimicrobial concentration that can be left in place for hours (41, 42). The data is based on anecdotal case reports primarily in infections due to CNS. Several groups have encouraging experience in salvaging long-term infected catheters with antibiotic lock in combination of systemic antimicrobial therapy (23). The antibiotic solution mixed with either heparin or saline may be installed in the infected catheter and left for the duration that catheter is not being used (over night). It is critical that this solution must be withdrawn prior to next infusion. We recommend that if antibiotic lock therapy is used in patients, systemic therapy must not be withheld, especially in immunosuppressed oncology patients, those with AIDS (43), granulocytopenia, and/or infection due to pathogen other than coagulase-negative staphylococci. Patients with extraluminal infection such as pocket or tunnel infection are not candidates for antibiotic lock therapy; systemic antimicrobial therapy along with catheter removal is the optimum approach in this setting. In select patients with both intraluminal and extraluminal infection with coagulase-negative staphylococci a trial of systemic antibiotics along with antibiotic lock approach may salvage the catheter. Due to high rate of success of antibiotic lock technique (44) in salvaging intravascular catheters we recommend that decision to retain and treat through an episode of CRBSI, antibiotic lock therapy may be entertained, especially in patients with history of difficult intravascular access.

COMPLICATED INFECTION

Patients that fail to resolve bloodstream infection after an infected intravascular device has been removed pose a serious diagnostic and treatment challenge. Persistent fever or bloodstream infection after 72 h following catheter removal must alert physician regarding possibility of complicated catheter-related infections such as, septic thrombophilibitis, endocarditis, osteomyelitis, or other metastatic foci of infection (23).

Endocarditis

Colonized intravascular catheters are the most common source of nosocomially acquired endocarditis, nearly 30% of patients with high-grade, persistent *S. aureus* bacteremia may have an underlying endovascular infection (45). Diagnostic approach includes TEE and radiographic work up for septic thrombus. Management includes prompt removal of intravascular catheter, and parentral antimicrobial therapy for 4 to 6 weeks. Surgical evaluation may be needed in the event of a large septic thrombus, and/or complications arising form endocarditis, such as severe valve dysfunction/damage, and valve ring and myocardial abscesses.

Septic Thrombophelibitis

Infected thrombus due to either bacterial or fungal infection involving peripheral vessel may often become clinically evident, however patients with great vein septic thrombosis may only present with fever, persistent/recurring bloodstream infection; whereas septic pulmonary embolism (cannon ball radiographic pattern) and/or paradoxical embolism are rare. *Staphylococcus aureus* is the most common organism, followed by *Candida* species and aerobic gram-negative bacilli. Ultrasonic examination alone may not provide diagnosis in all cases and select patients with high pre-test probability may need intravenous contrast-enhanced radiolograpic evaluation. Treatment of peripheral septic thrombophilibitis requires surgical incision and debridement and systemic antibiotics. For individuals with great vein thrombosis, antibiotic therapy is often prolonged (4 to 6 weeks) and anticoagulants may be given.

CONCLUSIONS

- Patients with short-term catheters: non-complicated CRBSI (non-persistent BSI, no evidence of sepsis) often respond to a short-course of parentral antibiotic therapy and removal of infected device.
- In non-complicated CRBSI involving surgically implanted tunneled catheters, catheter salvage approach may be used; systemic antimicrobial therapy (~ 14 days) may be given in conjunction with antibiotic lock therapy.

- In patients with complicated extraluminal (tunnel or pocket) infection catheter must be removed, and antibiotic therapy given for 10 to 14 days.
- In patients with CRBSI-associated septic thrombosis or endocarditis after prompt removal of infected device, antimicrobial therapy is continued for 4 to 6 weeks, and in the event of acute bone infection treatment is extended to 8 weeks.

REFERENCES

1.   Raad II. Intravascular-catheter-relted infections. Lancet 1998; 351: 893–898.
2.   Raad II, Hanna HA. Intravascular catherter-related infections. Arch Intern Med 2002; 162: 871–878.
3.   Mermel LA, Farr BM, Sheretz RJ, *et al*. Guidelines for the management of intravascular catheter-related infection. Clin Infect Dis 2001; 32: 1249–1272.
4.   Weinstein MP, Towns ML, Quartey SM, *et al*. The clinical significance of positive blood culture in the 1990s: a prospective comprehensive evaluation of the microbiology, epidemiology, and outcome of bacteeremia and fungemia in adults. Clin Infect Dis 1997; 24: 584–602.
5.   Warren DK, Zack JE, Elward AM, Cox MJ, Fraser VJ. Nosocomial primary bloodstream infections in intensive care unit patients in a nonteaching community medical center: a 21-month prospective study. Clin Infect Dis 2001; 33: 1329–1335.
6.   Edmond MB, Wallace SE, McClish DK, Pfaller MA, Jones RN, Wenzel RP. Nosocomial bloodstream infections in United States hospitals: a three-year analysis. Clin Infect Dis 1999; 29: 239–244.
7.   National Nosocomial Infections Surveillance (NNIS) System report, data summary from January 1990-May, 1999, issued June 1999. Am J Infect Control 1999; 27: 520–532.
8.   Diekema DJ, Pfaller MA, Schmitz FJ, *et al*. Survey of infections due to Staphylococcus species: frequency of occurrence and antimicrobial susceptibility of isolates collected fin he United States, Canada, Latin America, Europe, and the Western Pacific region for the SENTRY Antimicrobial Surveillance Program, 1997-1999. Clin Infect Dis 2001; 32(Suppl 2): S114–132.
9.   Costerton JW, Nickel JC, Ladd TI. Suitable methods for the comparative study of free-living and surface-associated bacterial population. Bact Nat 1986; 2: 49–84.

10. Calwell DE, Korber DR, Lawrence JR. Imaging of bacterial cells by fluorescence exclusion using scanning confocal laser microscopy. J Microbiol Meth 1992; 15: 249–261.

11. Sheretz RJ, Raad II, Balani A, Koo L, Rand K. Three-year experience with sonicated vascular catheter cultures in a clinical microbiology laboratory. J Clin Microbiol 1990; 28: 76–82.

12. Maki DG, Weise CE, Sarafin HW. A semiquantitative culture method for identifying intravenous catheter infections. N Engl J Med 1977; 296: 1305–1309.

13. Sherertz RJ, Carruth WA, Marosok RD, Espeland MA, Johnson RA, Solomon DD. Contribution of vascular catheter material to the pathogenesis of infection: the enhanced risk of silicone in vivo. J Biomed Mater Res 1995; 29: 635–645.

14. Deretic V, Schurr MJ, Boucher JC, Martin DW. Conversion of Pseudomonas aeruginosa to mucoldy in cystic fibrosis: environmental stress and regulation of bacterial virulence by alternative sigma factors. J Bacteriol 1994; 176: 2773 – 2780.

15. Farber BF, Kaplan MH, Clogston AG. Staphylococcus epidermidis extracted slime inhibits the antimicrobial action of glycopeptide antibiotics. J Infect Dis 1990; 161: 37–40.

16. Linares J, Sitges-Serra A, Garau J, Perez JL, Martin R. Pathogenesis of catheter sepsis: a prospective study with quantitative and semiquantitative cultures of the hub and segments. J Clin Microbiol 1985; 21: 357–360.

17. Maki DG. Infection caused by intravascular devices: pathogenesis, strategies for prevention. Royal Society of Medicine Services Ltd: London, 1991.

18. Kiehn TE, Armstrong D. Changes in the spectrum of organisms causing bacteremia and fungemia in immunocompromised patients due to venous access devices. Eur J Clin Microbiol Infect Dis 1990; 9: 869–872.

19. Raad II, Darouiche RO. Catheter-related septicemia: risk reduction. Infect Med 1996; 13: 807–812; 815–816; 823.

20. Kovacicova G, Lovaszova M, Hanzen J, Roidova A, Mateicka F, Lesay M, Krcmery V. Persistent fungemia—risk factors and outcome in 40 episodes. J Chemother 2001; 13: 429–433.

21. Farina C, Vailati F, Manisco A, Goglio A. Fungemia survey: a 10-year experience in Bergamo, Italy. Mycoses 1999; 42: 543–548.

22. Morrison VA, Weisdorf DJ. The spectrum of Malassezia infections in the bone marrow transplant population. Bone Marrow Transplant 2000; 26: 645–648.

23. Mermal LA, Farr BM, Sheretz RJ, et al. Guidelines for the management of intravascular catheter-related infections. Clin Infect Dis 2001; 32: 1249–1272.

24. Christensen GD, Bisno AL, Parisi JT, et al. Nosocomial septicemia due to multiple antibiotic-resistant Staphylococcus epidermidis. Ann Intern Med 1982; 96: 1–10.

25. Raad I, Davis S, Khan A, Tarrand J, Elting L, Bodey GP. Impact of central venous catheter removal on the recurrence of catheter-related coagulase-negative staphylococcal bacteremia. Infect Control Hosp Epidemiol 1992; 13: 215–221.

26. Lerner P, Weinstein L. Infective endocarditis in the antibiotic era. N Engl J Med 1966; 27: 388–393.

27. Rosen AB, Fowler VG, Corey GR, *et al.* Cost-effectiveness of transesophageal echocardiography to determine the duration of therapy for intravascular catheter-associated Staphylococcus aureus bacteremia. Ann Intern Med 1999; 130: 810–820.

28. Raad I. Optimal duration of therapy for catheter-related Staphylococcus aureus bacteremia: a study of 55 cases and review. Clin Infect Dis 1992; 14: 75–82.

29. Jernigan JA, Farr B. Short course therapy of catheter related Staphylococcus aureus bacteremia: a meta-analysis. Ann Intern Med 1993; 119: 304–311.

30. Durack DT, Lukes AS, Bright DK. New criteria for diagnosis of infective endocraditis: utilization of specific echocraiographic findings. Duke Endocarditis Service. Am J Med 1994; 96: 200–209.

31. Fowler VG, Sanders LL, Sexton DJ, *et al.* Outcome of Staphylococcus aureus bacteremia according to compliance with recommendations of infectious diseases speciealist: experience with 24 patients. Clin Infect Dis 1998; 27: 478–486.

32. Capdevila JA, Segarra A, Planes A, *et al.* Long term follow-up of patients with catheter related sepisi (CRS) treated without catheter removal [abstract J3]. In: Program and abstracts of the 35th Interscience Conference on Antimicrobial Agents and Chemotherapy (San Francisco). Washington, DC: American Society for Microbiology, 1995.

33. Schrenzel J, Schockmel G, Bregenzer T, *et al.* Severe staphylococcal infections: A randomized trial comparing quinolone + rifampin (iv then po) with conventional iv therapy [abstract 93]. In: Proceedings of the 36th annual meeting of the Infectious Diseases Society of America (San Francisco). Alexandria, VA: Infectious Diseases Society of America, 1998.

34. Vaudaux P, Francois P, Bisognano C, Schrenzel J, Lew DP. Comparison of levofloxacin, alatrofloxacin, and vamcomycin for prophylaxis and treatment of experimental foreign-body-associated infection by methacillin-resistant Staphylococcus aureus. Antimicrob Agents Chemother 2002; 46: 1503–1509.

35. Elting LS, Bodey GP. Septicemia due to Xanthomonas species and non-aeruginosa Pseudomonas species: increasing incidence of catheter-related infections. Medicine 1990; 60: 196–206.

36. Safdar A, Armstrong D. Infectious morbidity in critically ill patients with cancer. Critic Care Clin 2001; 17: 531–570.

37. Rex JH, Bennett JE, Sugar AM, *et al.* Intravascular catheter exchange and duration of candidemia. Clin Infect Dis 1995; 21: 994–996.

38. Rex JH, Bennett JE, Sugar AM, *et al.* A randomized trial comparing fluconazole with amphotericin B for the treatment of candidemia in patients without neutropenia. N Engl J Med 1994; 331: 1325–1330.

39. Mora_Duarte J, Betts R, Rotstein C, *et al.* Comparison of caspofungin and amphotericin B for invasive candidiasis. N Engl J Med 2002; 347: 2020–2029.

40. Arikan S, Lozano-Chiu M, Paetznick V, Rex JH. *In vitro* synergy of caspofungin and amphotericin B against *Aspergillus* and *Fusarium* spp. Antimicrob Agents Chemother 2002; 46: 245-247.

41. Gaillard JL, Merlino R, Pajot N, *et al.* Conventional and nonconventional modes of vancomycin administration to decontaminate the internal surface of catheters colonized with coagulase-negative staphylococci. JPEN Parenter Enteral Nutr 1990; 14: 593–597.

42. Douard MC, Arlet G, Leverger G, *et al.* Quantitative blood cultures for diagnosis and management of catheter-related sepsis in pediatric hematology and oncology patients. Intensive Care Med 1991; 17: 30–35.

43. Domingo P, Fontanet A, Sanchez F, *et al.* Morbidity associated with long-term use of totally implantable ports in patients with AIDS. Clin Infect Dis 1999; 29: 346–351.

44. Capdevila JA, Segarra A, Planes AM, *et al.* Successful treatment of hemodialysis catheter-related sepsis without catheter removal. Nephrol Dia Transplant 1993; 8: 231–234.

45. Fowler VG, Li J, Corey GR, *et al.* Role of echocardiography in evaluation of patients with Staphylococcus aureus bacteremia: experience in 103 patients. J Am Coll Cardiol 1997; 30: 1072–1078.

Chapter 8

THE MANAGEMENT AND TREATMENT OF INTRAVASCULAR CATHETER-RELATED INFECTIONS

Prof. T.S.J. Elliott
Department of Clinical Microbiology, University Hospital Birmingham NHS Trust, The Queen Elizabeth Hospital, Edgbaston, Birmingham, United Kingdom

Introduction

Catheter-related bloodstream infections continue to be associated with a significant morbidity and mortality. These infections are primarily related to the use of central intravascular catheters rather than peripheral devices (1) and the latter will therefore not be considered in this chapter.

In the US it has been estimated that more than 5 million central venous catheters (CVC) are used each year and that at least 400,000 cases of CVC-related infections occur annually. In comparison, in the UK over 10,000 episodes of CVC-related sepsis may occur each year (2-4).

Establishing the diagnosis of catheter-related sepsis relies both on clinical- and laboratory- based determinants. The clinical symptoms and signs of CVC sepsis are generally nondescript and usually await confirmation by laboratory findings including positive blood cultures (5). Clinical evidence suggesting intravascular device-related sepsis includes inflammation such as erythema and/or exudate at the catheter insertion site.

There may be no obvious source of infection in a patient commonly with a low grade pyrexia and non-specific symptoms including chills and rigors. Onset of symptoms such as transient pyrexia may also be linked to catheter manipulation or infusion administration. It is therefore evident that patients with a CVC *in-situ* need careful monitoring in order to make the diagnosis so that appropriate management can be commenced early in the infection.

MANAGEMENT OF CVC INFECTIONS

Two main factors need to be taken into account in the management of patients with CVC- related sepsis. Consideration has to be given as to whether or not the catheter needs to be removed and what is the most appropriate antimicrobial regimen to treat the infection (6).

CATHETER REMOVAL

It has been suggested that CVC should be routinely exchanged after 5 to 7 days *in-situ* through a guide wire and sent for culture even in patients without clinical evidence of sepsis. If the explanted catheter is subsequently shown to be colonized with microorganisms it may be necessary to replace the new catheter. The routine replacement of CVC in order to avoid sepsis is not however generally recommended (3).

In deciding whether or not to remove a CVC, when catheter-related sepsis is suspected, several factors need to be considered, these include:
- the patients' underlying condition and need for intravascular access.
- the type of CVC inserted in particular whether it is a short-term non tunneled catheter or a long-term surgically implanted catheter.
- the microbial cause of a CVC associated infection and its antimicrobial sensitivity pattern.
- the likelihood of successful replacement of a catheter at another site.
- the risk to the patient of removal and replacement.

Despite the clinical recognition that removal of a foreign body such as a colonized CVC is desirable to successfully treat device related infections, antimicrobials alone are being used in an attempt to salvage catheters particularly in patients with mild or moderate associated sepsis. However, if the patient fails to respond or deteriorates, the need for catheter removal should always be re-evaluated in view of any culture findings. If there is on-going sepsis with no resolution of clinical symptoms and signs and/or the cultures confirm the causative microorganism as virulent or difficult to treat for example *Candida albicans, Staphylococcus aureus,* coliforms or *Pseudomonas aeruginosa,* the catheter should be removed. With *S. aureus* CVC sepsis, there is substantial evidence that catheter removal results in the best outcome (7,8). Catheter removal for associated fungal infection has also decreased morbidity and mortality (9). Similarly with *P. aeruginosa* CVC sepsis, catheter removal has also improved patient outcome (10). Patients who develop an associated septic thrombosis or endocarditis also need catheter removal but antibiotic treatment should be continued for 4 to 6 weeks again depending on the causative microorganism.

The clinical requirement for catheter removal when associated with sepsis needs to be considered separately for non-tunnelled or tunnelled devices and these are presented below.

Microbiological investigations including blood cultures via a catheter and a separate venepuncture, skin entry site swabs for local sepsis and catheter tip examination on withdrawal of the device should be carried out where appropriate prior to commencement of antibiotics.

NON-TUNNELLED CVC

Non-tunneled central venous catheters associated with a mild to moderate infection, do not necessarily need to be routinely removed. Indeed, several studies (11,12), have demonstrated that the majority of catheters obtained from patients with only suspected catheter-related infection were sterile on removal suggesting that many devices were inappropriately explanted. It is generally accepted that the CVC should however be removed and examined microbiologically if the patient is severely unwell with a septicemia or if there is erythema and exudate at the catheter exit site (6). If the CVC is exchanged over a guide wire and significant colonization of the explanted catheter is subsequently demonstrated, the new catheter should also be removed and another device placed in a different site (13). When catheter tip cultures reveal significant growth without associated sepsis, the patient needs to be monitored closely. In this situation, if a significant pathogen such as *S. aureus* or *C. albicans* is isolated, a short course (5 to 7 days) of an appropriate intravenous antimicrobial is recommended by some authors (3).

If the causative microorganism of catheter-related sepsis is a coagulase negative staphylococcus such as *Staphylococcus epidermidis* and there is no evidence of local or systemic complications including entry site infection, septic shock or associated endocarditis, the catheter may be retained (14). However if the microorganism is more virulent or difficult to treat such as coliforms, *C.albicans* or *S. aureus* the catheter should ideally be removed. If removal of the device causes management difficulties such as patients with severe thrombocytopaenia or other coagulation problems, treatment with systemic antimicrobials and an antibiotic lock in the lumen of the catheter can be considered (15).

A summary of the main indications for the removal of non-tunneled CVC is given in Table 1.

Table 1. Indications for removal of non-tunnelled CVC.

- Skin site sepsis including erythema and purulent exudate

- Associated septicemia or bacteraemia

- Specific infections including *S.aureus, Candida* spp. or coliforms

- CVC exchanged over guide wire and associated with significant colonization

- Evidence of metastatic infection

- Deteriorating patient condition

- Treatment failure to antibiotics

TUNNELLED CVC

Tunneled CVC are implantable devices, which include those which are surgically inserted including Hickman catheters or subcutaneous infusion ports such as a port-a-cath. With tunneled devices, it is particularly important to clinically establish whether or not a patient has a catheter-related infection rather than colonization. This is because it can be relatively difficult to remove a surgically implanted device. Blood for culture taken under aseptic conditions via the device as well as a separate peripheral venepuncture should therefore be obtained. This should include obtaining the appropriate volume of blood to inoculate into the culture bottles to facilitate isolation of any causative microorganism and to allow direct comparison of time to positivity of the different blood cultures. Patients with a confirmed tunnelled CVC associated infection usually require removal of the catheter and up to 10 days of antibiotic therapy depending on the causative microorganism. For tunneled catheter-related sepsis including septicemia caused by more virulent microorganisms or those which are more difficult to treat including *S. aureus*, *Candida spp.* or coliforms, the devices should ideally be removed. Attempts to salvage tunneled devices may be made with systemic antimicrobial and antibiotic lock therapy but this approach is dependent on the patients clinical condition and the underlying pathogen. In comparison, in the presence of uncomplicated infection caused by low virulent pathogens such as the coagulase negative staphylococci, the device may be retained if there is no evidence of local infection at the insertion site or continuing bacteremia or septicemia (16).

ANTIMICROBIAL TREATMENT FOR CATHETER-RELATED SEPSIS

Antibiotic Lock

An antibiotic lock involves the administration of antibiotics which remain in the internal lumen of the catheter for an extended period of time. Intraluminal therapy has the advantages of directing the antimicrobial to the focus of microbial colonization and infection and it can be administered in the out-patient situation (17). It has been clearly demonstrated for example that it is possible to successfully kill susceptible microorganisms attached to the internal lumen of catheters using high concentrations of vancomycin locked within the device. A number of clinical reports have also subsequently reported high rates of successful therapy with use of the antimicrobial lock technique, Messing et al. (18) for example used an antibiotic lock for 12 hours per day for up to 16 days together with a 3 day course of systemic antibiotics for patients with catheter-related sepsis. No patients had infection at the insertion site nor a tunnel infection. The infections were controlled without catheter removal in 20 out of 22 patients. Two treatment failures occurred and were associated with candida sepsis.

The likelihood of a CVC related infection being successfully treated with antibiotics locked in the catheter is dependent on several factors including the type of infection and the causative microorganism. It is for example relatively easy to treat an exit site infection rather than a tunnel infection. Infections associated with coagulase negative staphylococci are more likely to respond than sepsis related to S.aureus or *P. aeruginosa* (3,7). Several open trials of antibiotic lock therapy with or without parenteral treatment used for tunneled catheter-related bacteremias have reported an overall response and catheter salvage rate of over 80% (138 in 167 episodes). This compares to the standard parenteral therapy for the treatment of tunneled catheter-related bloodstream infections with a salvage rate of 66.5% (342 in 514 episodes) (3,7). The use of the antibiotic lock therefore appears to be significantly more effective in treating CVC related sepsis resulting in catheter salvage. However, fungal infections poorly respond to anti-fungal lock therapy (3,19,20) and catheter removal should always be considered in this group of patients.

Antibiotic lock therapy for catheter-related bacteremia or septicemias is now often used in conjunction with systemic antibiotic therapy. It is however important that in patients with moderate or severe infections including catheter entry site infections, tunnel infections or septicemia, that systemic antibiotics are given in addition to an antibiotic lock (21). The concentration of the antibiotic in the lock, must be sufficient enough to penetrate the biofilm in which the bacteria adhere to the surface of the internal catheter lumen. The antibiotics used in the lock are usually given in a concentration of between 1 to 5 mg/ml and mixed with a heparin flush to completely fill the internal lumen catheter which, dependent on the device used, is approximately 2ml. Vancomycin has been used at concentrations of 1 to 5 mg/ml, gentamicin at 1 to 2 mg/ml and ciprofloxacin at 1 to 2 mg/ml (3). The volume of instilled antibiotic is usually removed before the next use of the CVC such as when giving intravenous medication. The length of antibiotic lock treatment is unclear and has varied in the various studies but has been on average about 2 weeks (3).

Systemic Antimicrobial Therapy

The commonest microorganisms which cause catheter-related bloodstream infections include the coagulase negative staphylococci and *S. aureus*. Catheter-related bloodstream infections should be treated with parenteral antibiotic therapy at full therapeutic dose where appropriate. Treatment should ideally be administered via the presumed colonized catheter which will remain a focus of infection if retained. The choice of antimicrobial agent should be guided by the antibiotic susceptibility of the causative microorganism if known. However, in severely ill patients, empirical antibiotic treatment needs to be started based on local susceptibility patterns and the likely infecting microorganism. Empirical antibiotic treatment is required if the catheter is not removed; if the patient has moderate to severe sepsis including septic shock; if the patient is immunosuppressed or has an indwelling prosthesis such as a replacement heart valve (6). Empirical antibiotic treatment has also been recommended for patients with prosthetic joints (22).

SPECIFIC ANTIMICROBIAL THERAPY

Glycopeptides such as vancomycin and teicoplanin are active against most staphylococci including the methicillin resistant coagulase negative staphylococci and *S. aureus*. There have been a number of studies which have compared vancomycin and teicoplanin for the treatment of CVC sepsis and they have shown similar efficacy with high response rates, (23,24). Other trials have reported reduced efficacy with teicoplanin as compared to vancomycin to treat staphylococcal sepsis (25). This may reflect resistance of some coagulase negative staphylococci, for example *Staphylococcus haemolyticus* to teicoplanin. Empirical treatment with vancomycin may therefore be more appropriate than teicoplanin, if the patient has an infection caused by a staphylococcus resistant to flucloxacillin.

Gram negative aerobic bacilli primarily cause between 10 and 20% of all episodes of CVC related sepsis. Therefore, in patients exhibiting clinical evidence of Gram negative aerobic sepsis, additional empirical cover for coliforms and also for *P.aeruginosa* needs to be considered, particularly in hospitalised patients. This includes the considering the use of intravenous cephalosporins such as ceftazidime, the quinolones for example ciprofloxacin or an aminoglycoside including gentamicin. The choice of antimicrobial must be guided by the antimicrobial local sensitivity pattern of pathogens and what type of microorganisms are predominant in in any particular clinical area.

If the patient has a suspected fungaemia, then intravenous fluconazole should be considered (26). However, if the patient is colonized or the infection is caused by *Candida* spp. species other than *Candida albicans*, then amphotericin should be considered particularly in the seriously ill patient. Anti-fungal therapy should be also be considered for patients with previous CVC related candidemia. Many patients with fungal CVC related infections have been treated with amphotericin and together with removal of the catheter this has resulted in a good response. It is unclear however, as to the length of therapy required (26). It is usually recommended that antifungal treatment is given for at least 14 days after the last positive culture.

There are very few controlled trials with adequate numbers of patients to confirm the optimal antibiotics and their duration for the treatment of catheter-related bloodstream infection. Indeed further research is required to answer the question of optimal length of treatment. Conventional recommendations advise prolonged treatment if there is concern of underlying associated infections such as septic thrombosis, osteomyelitis or endocarditis. However, other studies have suggested that shorter courses (10-14 days) of treatment may be given if the risk of infection related complications is relatively low (8). Many of these studies have been only on a limited number of patients making analysis difficult. Conversely (27), in a meta-analysis of published studies it was concluded that short courses of antibiotics for uncomplicated catheter-related *S.aureus* bacteremia could not be supported. Despite this, substantial numbers of patients receiving short courses of antimicrobial therapy do respond adequately. It is therefore generally accepted that if there is a relatively rapid response to the initial antibiotic treatment and for patients who are not immunocompromised without an underlying associated clinical risk factor such as a prosthetic device *in situ*, then they should receive between 10 to 14 days of antimicrobial therapy.

Up to one third of patients with CVC related sepsis develop a major complication including septic shock, phlebitis, metastatic infection and endocarditis and this should always be taken into account (Arnow 1993). This underlines the importance of identifying and treating these infections early and with the most appropriate antimicrobial agent. Persisting bacteraemia, septicaemia or fungaemia after catheter removal and commencement of appropriate antimicrobial treatment may suggest a metastatic infection. If there is evidence of associated infection related to the bacteraemia including septic thrombosis or endocarditis, longer courses of therapy should always be considered, up to 6 weeks for endocarditis or septic thrombosis and up to 8 weeks for osteomyelitis.

RECOMMENDATIONS FOR THE MANAGEMENT OF SPECIFIC INFECTIONS

Blood cultures and swabs of infected insertion sites should be taken prior to commencement of the antimicrobial therapy. When a causative microorganism has been identified and its sensitivity pattern determined, then specific directed antimicrobial therapy can be commenced.

Table 2 summarizes the antibiotics which may be used for the different causative microorganisms of CVC sepsis. The antimicrobial treatment of selected specific pathogens is dealt with in more detail below.

Vancomycin Resistant Staphylococci or Enterococci

Quinupristin-dalfdopristin, a semi-synthetic streptogramin, is an alternative to glycopeptides for the treatment of catheter-related infections caused by vancomycin resistant enterococci or staphylococci. In a randomized trial involving 39 patients with catheter-related staphylococcal bacteraemia quinupristin-dalfopristin was similar in efficacy to vancomycin (28).

Table 2. Intravenous Antimicrobial Treatment of Central Venous Catheter-Related Bloodstream Infection in Adults.

Causative microorganism	+ Antimicrobial
Gram-positive cocci	
S.aureus	
Methicillin sensitive	Flucloxacillin
Methicillin resistant	Vancomycin
Coagulase-negative staphylococci	
Methicillin sensitive	Flucloxacillin
Methicillin resistant	Vancomycin
E.faecalis/E.faecium	
Ampicillin sensitive	Ampicillin ± aminoglycoside
Ampicillin resistant, vancomycin sensitive	Vancomycin ± aminoglycoside
	Linezolid or Quinupristin/Dalfopristin
Vancomycin resistant	
Gram-negative bacilli	
Coliforms e.g. *E.coli* and *Klebsiella* species	Cefotaxime or Ciprofloxacin
Pseudomonas aeruginosa	Ceftazidime ± aminoglycoside
Fungi	
C.albicans or Candida species (non albicans)	Amphotericin or fluconazole (if *Candida* spp. is susceptible)

+ the antimicrobial agent recommended needs to reflect the sensitivity pattern of the microorganism causing the sepsis and may therefore need modification when culture and sensitivity results are available.

Staphylococcus aureus

As discussed above, if a nontunnelled CVC is suspected to be the source of *S.aureus* bacteraemia, the device should be removed, and a new catheter should be re-inserted at a different site (8,19,29). Similarly, tunnelled CVC should be removed if there is evidence of tunnel, or exit-site infection (3,7,30). Patients with uncomplicated catheter-related bacteremia or septicemia caused by *S. aureus* should be treated for a minimum of 10 to 14 days after catheter removal.

When catheter salvage is clinically critical and a decision is made to retain the device, treatment with appropriate systemic antimicrobial and antibiotic lock therapy for at least 14 days may be attempted (3,31).

However in a report of 50 patients with *S. aureus* catheter-related sepsis, delayed catheter removal was associated with increased mortality (8). This needs to be taken into account when attempting catheter salvage. Similarly, in a meta-analysis of 11 studies, 8 of 132 patients, who had a short course of treatment for *S. aureus*, developed endocarditis or infection at another site (27).

Echocardiography should therefore be carried out on at risk patients to diagnose possible endocarditis that requires therapy for 4-6 weeks (19,32). If there is evidence of other distant foci of infection, such as septic venous thrombosis, treatment should similarly be continued for several weeks. Patients with a negative echocardiogram whose catheter has been removed should be treated with appropriate antibiotics for 10-14 days with systemic antibiotic therapy (19,32).

Coagulase-Negative Staphylococci

Vancomycin or flucloxacillin for 7 days should be used to treat CVC-related sepsis caused by coagulase negative staphylococci, if the isolate is susceptible (3,33,34) and the catheter removed. If the CVC is retained (non-tunneled or tunneled), systemic antibiotic therapy together with antibiotic lock therapy for up to 14 days should be considered. (3,17,19-21).

CONCLUSIONS

CVC related sepsis offers a challenge to the clinician both in terms of diagnosis and treatment. The general therapeutic approaches include removal of the device with a short course of appropriate antibiotics. Removal of a CVC is desirable particularly with implanted devices and difficult to treat infections which may metastasize including *S. aureus*, coliforms or *Candida spp*. If a device is kept *in-situ* and is associated with sepsis, treatment with systemic and line locked antimicrobials needs to be considered, but may not be successful. Infectious complications including endocarditis and septic thrombosis associated with CVC are relatively common and patients should always be monitored for these conditions.

REFERENCES

1.   Fletcher SJ, Bodenham AR. Catheter-related sepsis-an overview-Part 1. Br J Intensive Care 1999;9:46-53.
2.   Elliott TSJ. The prevention of central venous catheter-related sepsis. J Chemother 2001;13 Spec No 1:234-238.
3.   Mermel LA, Farr BM, Sherertz RJ, Raad, II, O'Grady N, Harris JS, Craven DE. Guidelines for the management of intravascular catheter-related infections. Clin Infect Dis 2001;32:1249-1272.
4.   Berrington A, Gould FK. Use of antibiotic locks to treat colonized central venous catheters. J Antimicrob Chemother 2001;48:597-603.
5.   Elliott TSJ, Roth JR. Characterization of Tn10d-Cam: a transposition-defective Tn10 specifying chloramphenicol resistance. Mol Gen Genet 1988;213:332-338.
6.   Rodriguez-Bano J. Selection of empiric therapy in patients with catheter-related infections. Clin Microbiol Infect 2002;8:275-281.
7.   Dugdale DC, Ramsey PG. Staphylococcus aureus bacteremia in patients with Hickman catheters. Am J Med 1990;89:137-141.
8.   Malanoski GJ, Samore MH, Pefanis A, Karchmer AW. *Staphylococcus aureus* catheter-associated bacteremia. Minimal effective therapy and unusual infectious complications associated with arterial sheath catheters. Arch Intern Med 1995;155:1161-1166.
9.   Nguyen MH, Peacock JE, Jr., Tanner DC, Morris AJ, Nguyen ML, Snydman DR, Wagener MM, Yu VL. Therapeutic approaches in patients with candidemia. Evaluation in a multicenter, prospective, observational study. Arch Intern Med 1995;155:2429-2435.
10.   Kuikka A, Valtonen VV. Factors associated with improved outcome of *Pseudomonas aeruginosa* bacteraemia in a Finnish university hospital. Eur J Clin Microbiol Infect Dis 1998;17:701-708.
11.   Rello J, Coll P, Prats G. Evaluation of culture techniques for diagnosis of catheter-related sepsis in critically ill patients. Eur J Clin Microbiol Infect Dis 1992;11:1192-1193.
12.   Henderson DK. Bacteremia due to percutaneous intravascular devices. In: Mandell GL, Bennett JE, Dolin R, editors. Principles and Practice of Infectious Diseases. 4th ed. New York: Churchill-Livingstone; 1995.
13.   Pettigrew RA, Lang SD, Haydock DA, Parry BR, Bremner DA, Hill GL. Catheter-related sepsis in patients on intravenous nutrition: a prospective study of quantitative catheter cultures and guidewire changes for suspected sepsis. Br J Surg 1985;72:52-55.
14.   Raad I, Davis S, Khan A, Tarrand J, Elting L, Bodey GP. Impact of central venous catheter removal on the recurrence of catheter-related coagulase-negative staphylococcal bacteremia. Infect Control Hosp Epidemiol 1992;13:215-221.
15.   Raad I. Intravascular-catheter-related infections. Lancet 1998;351:893-898.

16. Raad, II, Sabbagh MF. Optimal duration of therapy for catheter-related bacteremia: a study of 55 cases and review. Clin Infect Dis 1992;14:75-82.

17. Gaillard JL, Merlino R, Pajot N, Goulet O, Fauchere JL, Ricour C, Veron M. Conventional and nonconventional modes of vancomycin administration to decontaminate the internal surface of catheters colonized with coagulase-negative staphylococci. JPEN J Parenter Enteral Nutr 1990;14:593-597.

18. Messing B, Peitra-Cohen S, Debure A, Beliah M, Bernier JJ. Antibiotic-lock technique: a new approach to optimal therapy for catheter-related sepsis in home-parenteral nutrition patients. JPEN J Parenter Enteral Nutr 1988;12:185-189.

19. Benoit JL, Carandang G, Sitrin M, Arnow PM. Intraluminal antibiotic treatment of central venous catheter infections in patients receiving parenteral nutrition at home. Clin Infect Dis 1995;21:1286-1288.

20. Krzywda EA, Andris DA, Edmiston CE, Jr., Quebbeman EJ. Treatment of Hickman catheter sepsis using antibiotic lock technique. Infect Control Hosp Epidemiol 1995;16:596-598.

21. Rao JS, O'Meara A, Harvey T, Breatnach F. A new approach to the management of Broviac catheter infection. J Hosp Infect 1992;22:109-116.

22. Capdevila JA. Catheter-related infection: an update on diagnosis, treatment, and prevention. Int J Infect Dis 1998;2:230-236.

23. Chow AW, Jewesson PJ, Kureishi A, Phillips GL. Teicoplanin versus vancomycin in the empirical treatment of febrile neutropenic patients. Eur J Haematol Suppl 1993;54:18-24.

24. Menichetti F, Martino P, Bucaneve G, Gentile G, D'Antonio D, Liso V, Ricci P, Nosari AM, Buelli M, Carotenuto M, *et al.* Effects of teicoplanin and those of vancomycin in initial empirical antibiotic regimen for febrile, neutropaenic patients with haematologic malignancies. Gimema Infection Program. Antimicrob Agents Chemother 1994;38:2041-2046.

25. Kaatz GW, Seo SM, Dorman NJ, Lerner SA. Emergence of teicoplanin resistance during therapy of *Staphylococcus aureus* endocarditis. J Infect Dis 1990;162:103-108.

26. Rex JH, Bennett JE, Sugar AM, Pappas PG, van der Horst CM, Edwards JE, Washburn RG, Scheld WM, Karchmer AW, Dine AP, *et al.*. A randomized trial comparing fluconazole with amphotericin B for the treatment of candidemia in patients without neutropenia. Candidaemia Study Group and the National Institute. N Engl J Med 1994;331:1325-1330.

27. Jernigan JA, Farr BM. Short-course therapy of catheter-related *Staphylococcus aureus* bacteremia: a meta-analysis. Ann Intern Med 1993;119:304-311.

28. Raad I, Bompart F, Hachem R. Prospective, randomized dose-ranging open phase II pilot study of quinupristin/dalfopristin versus vancomycin in the treatment of catheter-related staphylococcal bacteremia. Eur J Clin Microbiol Infect Dis 1999;18:199-202.

29. Libman H, Arbeit RD. Complications associated with *Staphylococcus aureus* bacteremia. Arch Intern Med 1984;144:541-545.

30. Benezra D, Kiehn TE, Gold JW, Brown AE, Turnbull AD, Armstrong D. Prospective study of infections in indwelling central venous catheters using quantitative blood cultures. Am J Med 1988;85:495-498.

31.   Capdevila JA, Segarra A, Planes A. Long term follow-up of patients with catheter
        related sepsis (CRS) treated without catheter removal. [abstract 53]. In: Program and
        abstracts of the 35th Interscience conference on Antimicrobial and Chemotherapy (San
        Francisco). Washington, D.C.: American Society for Microbiology 1995.
32.   Johnson DC, Johnson FL, Goldman S. Preliminary results treating persistent central
        venous catheter infections with the antibiotic lock technique in pediatric patients.
        Pediatr Infect Dis J 1994;13:930-93.
33.   Chambers HF, Miller RT, Newman MD. Right-sided *Staphylococcus aureus*
        endocarditis in intravenous drug abusers: two-week combination therapy. Ann Intern
        Med 1988;109:619-624.
34.   Archer GL. *Staphylococcus epidermidis* and other coagulase negative staphyococci.
        In: Mandell GL, Bennett JE, Dolin R, editors. Principles and Practice of Infectious
        Diseases. 5th ed. New York: Churchill Livingstone; 2000. p 2092-2100.

Chapter 9

EDUCATION AS THE PRIMARY TOOL FOR PREVENTION

Phillippe Eggimann, M.D., Didier Pittet, M.D., M.S.
Medical Intensive Care Unit and Infection Control Program, Department of Internal Medicine, University of Geneva Hospitals, Geneva, Switzerland

Introduction

The prevention of catheter-associated infections relies above all on a strict observance of the basic rules of hygiene. Among these, hand hygiene can be considered as the first and most important. More specific measures, including the use of maximal sterile barriers during insertion, optimal insertion site preparation, detailed guidelines for catheter replacement, and defining particular situations for the use of antiseptic/antibiotic-coated devices have been studied in detail in many clinical studies. Detailed guidelines for the insertion and the care of vascular accesses are regularly published, but data from surveillance programs have shown repeatedly that they are generally either not applied or inadequately so.

GUIDELINES

Thousands of guidelines have been proposed to physicians. Many of them concern preventive care (1,2). They represent a major tool to identify the best evidence-based procedure of care to educate healthcare workers (HCWs) and to improve the quality of healthcare delivery (3).

However, many physicians remain wary of their intent. A survey of American College of Physicians members published in 1994 indicated that 43% of physicians believed that guidelines would increase healthcare costs, 68% that they would be used to discipline them, and 34% that they would make medical practice less satisfying (4). Although evidence has suggested that guidelines could improve both the process and the outcome of patient care, the degree of improvement has varied and may only have been transient (2). Accordingly, efficient guideline implementation requires their integration in a manner that effectively communicates best practice to be incorporated by HCWs (5-7).

GENERAL MEASURES

As for any other nosocomial infection, the prevention of vascular access-related infections relies on a strict respect of the basic rules of hygiene, particularly hand hygiene procedures.

It has now been clearly and firmly established that the promotion of hand rubbing with alcohol-based formulations may result in significant and prolonged improvement of hand hygiene practices compared to traditional hand washing with soap and water, for which compliance rarely exceeds 40% (8). Hand rubbing combines the advantages of a rapid action with more potent antimicrobial efficacy at a lower cost. Accordingly, guidelines for hand hygiene procedures have been completely reviewed and adapted to these concepts (9).

TRAINING AND EDUCATIONAL PROGRAMS

Training and educational programs specifically designed to reduce the incidence of catheter-associated infections were recently proved to be effective (10-12). Education of the HCWs in charge of the insertion and handling of vascular access in ICUs, where almost all patients are equipped with at least one IV line, was the cornerstone of these programs. Their objective was to obtain adherence to previously published guidelines and standardization of care at the bedside as established by ward reference staff in close collaboration with infection control specialists.

THE NORTH CAROLINA EXPERIENCE

Sherertz *et al.* (10) recently reported that an educational program of physician-in-training can decrease the risk of catheter-related infections. They reported on 3090 catheters inserted over an 18-month period in six ICUs and in one step-down unit after the introduction of the program.

The program consisted of a one-day course on infection control practices and vascular access insertion procedures. It included a one-hour introduction of basic infection control principles (hand hygiene, isolation and barrier use, handling of patients with resistant organisms and varicella). Thereafter, students and physicians rotated through a series of one-hour stations, during which they received 5 to15 minutes of didactic instruction followed by hands-on instruction overseen by faculty members. Training was provided in: 1) blood-draws through vascular lines; 2) arterial puncture; 3) insertion of arterial lines and central venous catheters (CVCs); 4) urinary catheter insertion; 5) lumbar puncture; 6) peripheral venous catheter insertion; and 7) phlebotomy. Participants were also instructed to change dressings and intravenous tubing every three days and not to adhere to fixed schedules for changing CVCs (Table 1).

This program was shown to reduce the associated infection rate by 28%, from 3.3 to 2.4 episodes/1000 CVC-days.

THE GENEVA EXPERIENCE

We conducted a study to evaluate the impact of a global strategy targeted at the reduction of catheter-related infections in 3154 critically-ill patients consecutively admitted to a medical ICU (11).

The program consisted of slide-show-based educational sessions and bedside training of both physicians and nurses and included specific recommendations for the insertion and handling of vascular accesses (13,14) (Table 2).

Following the introduction of this educational program, the incidence-density of exit-site catheter infection decreased by 64%, and that of primary bloodstream infections by 67% (11). Although the overall exposure to CVCs did not significantly differ between the control and the intervention periods (median duration, 4 days, P=0.94), the incidence-density of bloodstream infections markedly decreased from 22.9 to 6.2 episodes/1000 CVC-days due to a reduced incidence of both microbiologically-documented infection (from 6.6 to 2.3 episodes/1000 CVC-days) and clinical sepsis (from 16.3 to 3.9 episodes/1000 CVC-days). Overall, the incidence-density of all ICU-acquired nosocomial infections was reduced by 35% (from 52.4 to 34.0 episodes/1000 patient-days). This corresponded to the prevention of 50 to 104 nosocomial infections over an 8-month period including at least 1 to 11 primary bloodstream infections, 15 to 29 clinical sepsis, and 15 to 32 vascular-access related infections.

Table 1. Detailed Content of a 1-Day "Hands-On" Course on Basic Procedures and Infection Control Practices to Physicians-In-Training Newly Incorporated in the Staff of Six ICUs and One Step-Down Unit*.

1-hour course [†]	**Basic infection control principles**
	-Hand hygiene procedures
	-Appropriate use of barrier garments
	-Handling of patients with resistant organisms
1-hour station [‡]	**Training stations**
	-Blood-draw through vascular lines
	-Arterial puncture
	-Insertion of arterial lines and central venous catheters
	-Urinary catheter insertion
	-Lumbar puncture
	-Peripheral venous catheter insertion [&]
	-Phlebotomy[&]
Guidelines	**Recommendations targeted at vascular access care**
	-Use of povidone-iodine for skin preparation
	-Use of full-size sterile drapes for insertion
	-Avoidance of antibiotic ointment at the insertion site
	-Use of clear plastic dressings
	-Dressing and tubing change every 3 days
	-No fixed schedules for changing central venous catheters

* Adapted from reference (10)

† Completed by a 1-hour course on Occupational Safety and Health Administration (http://www.osha.gov) for blood and body fluids and for tuberculosis on a different day ; ‡ Consisting o f 5-15 minutes of didactic instruction, followed by hands-on training on mannequins

& Training was performed on mannequins, then on other participants

Table 2. Detailed Guidelines for Insertion and Handling of Vascular Accesses to Prevent Catheter-Related Infections, Medical ICU, University of Geneva Hospitals*.

Hygiene	Hand disinfection:	Strongly emphasized for any care (http://www.hopisafe.ch)
	Hand washing:	Restricted to visibly soiled hands, followed by hand disinfection
Material	Preparation:	Material disposed according to detailed listing to avoid interruptions during insertion [†]
Patient	Installation:	Patient and devices placed to provide sufficient access to the insertion site for the operator
Insertion	Skin preparation:	Hair-cutting instead of shaving
	Antisepsis:	Alcohol-based (75%, v/v) solution with chlorhexidine gluconate (0.5%)
	Technique:	Maximal barrier precautions: sterile gown and gloves, cap, surgical mask, large sterile drapes
	Site:	Promotion of subclavian (CVC) and wrist vein (short lines) sites
	Fixation:	Promotion of simple node at the exit-site, without special fixing device
Dressing	Transparent dress:	Occlusive devices without gauze not allowed
	Dry gauze:	Occlusion with porous adhesive band imposed
Handling	General measure:	New caps after any opening of the hubs
	Blood sampling:	On antiseptic-impregnated pads
	Drug infusions:	On antiseptic-impregnated pads; new temporary pipe for each administration
	Cardiac output:	Closed system only, without opening of the circuit
Replacement	72 hr intervals:	For dressings, sets, pipes and devices
	24 hr intervals:	For lipid or blood product lines
Removal	In general:	Peripheral lines after 72 hrs
		Central lines as clinically indicated
		Prompt removal if vascular accesses not absolutely necessary
	Special conditions:	Guidewire exchange systematically performed for any unexplained clinical sepsis [‡]

* Adapted from references (11,13).

† Precise listing of the material needed as well as detailed description of the insertion process must be given to all the staff of the unit including physicians, nurses and nursing assistants

‡ Clinical sepsis was defined as the one of the following clinical signs or symptom with no other recognized cause: fever (>38°C), hypotension (systolic blood pressure £ 90 mmHg), or oliguria (<20 ml/hr) and all of the following: blood culture not performed; no apparent. infection at another site; physician institutes appropriate antimicrobial therapy for sepsis (14).

Table 3. Detailed Content of the Education Program for Insertion and Maintenance of Central Venous Catheters Directed Toward Registered ICU Nurses to Reduce Catheter-Related Infections Rates (adapted from reference 12).

Self-study module	**Information related to catheter-related infections**
	 1. Epidemiology and scope of the problem
	 2. Risk factors, with special emphasis on
	    - length of hospitalization and of catheterization time
	    - colonization of insertion site and hub
	    - anatomical location of central venous catheter insertion
	 3. Etiology
	 4. Definitions
	 5. Methods to decrease risk
	    - hand washing and antisepsis technique
	    - methods for detecting clinical signs and symptoms of local infection
	    - technique for sending catheter-tip culture
	    - routine catheter site care
	    - replacing administration sets and fluids
	    - cleaning and changing injection ports and luer-lock caps
	    - how to handle parenteral fluids and multidose vials
	    - procedure for drawing blood cultures

Guidelines	**Guidelines for catheter maintenance** - Changing injection caps and intravenous tubing for fluids and medications every 72 hrs (or immediately if blood accumulated in or round the cap or its integrity was compromised) - Replacement of transparent line dressings every 7 days - Replacement of gauze dressings, used solely when there was bleeding or oozing at the insertion site, every 48 hrs - Immediate replacement of dressings that were soiled or no longer occlusive
Evaluation	**Pre- and postintervention test** - Pre and post-test and the study module were mandatory for all registered nurses in the ICU - Nurses who scored below 80% correct on the post-test were required to repeat the module **Information covered on pre- and postintervention test** - Number of hospital-acquired bloodstream infections in the US/year - Percentage of ICU bacteremias associated with intravenous devices - Aseptic technique - Risk of infection in subclavian central venous catheters - Pathophysiology of contamination of catheter hub - Clinical manifestations of catheter-related bloodstream infection - Sending catheter tip for culture - Technique and timing of scheduled dressing changes - Handwashing, sterile glove and mask usage - Antimicrobial ointment usage - Frequency of intravenous tubing and luer-lock cap changes - Timing and dating of multidose vials - Accessing central venous catheter ports - Blood culture technique

THE ST. LOUIS EXPERIENCE

More specifically designed to improve the handling of vascular accesses by the nursing staff, an education-based program was introduced in a surgical ICU at the Barnes-Jewish hospital in St Louis, Missouri (12). Based on a self-learning process, staff were asked to answer a series of questions
before and after reading a brochure on catheter-related infections. The brochure included epidemiological data, physiopathological concepts and detailed preventive measures. Key recommendations were supported by posters and regular reminders in the ward, similar to the program described in Geneva (11,15) (Table 3).

Here again, the impact was very impressive. Eighteen months following the introduction of the program, the incidence-density of primary bacteremia was reduced from 10.8 to 3.7 episodes/1000 catheter-days (12).

COST EFFECTIVENESS OF EDUCATIONAL PROGRAMS

Although the attributable costs of vascular-access related infections remain to be precisely determined, the overall benefit from the described education programs can be assessed. Using a conservative approach to estimate resource use, the reduction in nosocomial infections reported after the introduction of these programs was at least as effective as the reduction that could be expected if antiseptic-coated catheters would have been introduced in the corresponding wards.

Sherertz *et al.* estimated that the introduction of the program was followed by cost savings of between US$ 63,000-800,000 (10). This corresponded to the salary of a specialized nurse in all six participating units for one month in the low hypothesis to one year in the high hypothesis. According to the cost-efficacy analysis conducted by Veenstra *et al.* which estimated that the use of each chlorhexidine-silver-sulfadiazine coated-catheter saved US$ 68-391 (16), this may correspond to the anticipated benefit gained with the use of 161 to 926 devices in the lower hypothesis and of 2035 to 11,767 in the high hypothesis, as compared to an average of 390 catheters used during the study period.

Cost savings one year after the intervention in Geneva would correspond to the annual salary of 3 to 4 full-time infection control nurses. According to the cost-efficacy analysis performed by Veenstra *et al.* (16), this may also correspond to the anticipated benefit gained with 540 to 3103 catheters for the lower benefit hypothesis and from 3061 to 6102 catheters for the upper hypothesis, as compared to an average of 500 catheters annually used in the unit.

In the experience from St Louis, by multiplying the estimated cost resulting from the number of infections avoided (US $3700-24,000), Coopersmith *et al.* evaluated that their program saved US $185,000-2,808,000 over an 18-month period. This corresponds to 2 to 30 specialized nurses or between 473 to 41,294 coated catheters, as compared to 727 catheters placed during the same period.

LONG-TERM IMPACT

Preliminary results from Geneva, 30 months after the implementation of the prevention program, showed that catheter-related infection rates decreased by a further 25% as compared to the results from the initial report. This represents a 90% decrease compared to the initial rates (17) and strongly suggests that the integration in daily practice of most of the elements included in the preventive program is effectively possible.

NEW RECOMMENDATIONS

These observations indicate that behavioral changes may have played a key role in the success of the described educational programs which were based on multimodal and multidisciplinary approaches including communication and education tools, active participation of HCWs and positive feedback, and systematic involvement of institutional leaders (18,19).

Accordingly, the concept of preventing catheter-related infections has evolved, and education-based programs, including some adaptation of specific measures by the staff of the targeted wards, are now recommended as a first line strategy in the recently updated guidelines for the prevention of these infections (7).

CONCLUSIONS

In the absence of other clinical focus of infection, vascular accesses are responsible for the majority of primary bloodstream infections and clinical sepsis. Potentially lethal, these infections account for a significant prolongation of hospital stay, additional morbidity, and associated costs. However, a large number of these infections are preventable, and they should no longer be viewed as an inexorable tribute to pay to technologic medicine.

Their prevention should be conceived as improvement of the quality of care and based on the introduction of education-based programs. These programs should include both general measures targeted at an improved observance of the basic hygienic rules, such as hand hygiene procedures, and more specific measures to be progressively included in the standard of care.

REFERENCES

1.   Tunis SR, Hayward RS, Wilson MC, Rubin HR, Bass EB, Johnston M, Steinberg EP. Internists' attitudes about clinical practice guidelines. Ann Intern Med 1994;120:956-63.
2.   Grimshaw JM, Russell IT. Effect of clinical guidelines on medical practice: a systematic review of rigorous evaluations. Lancet 1993;342:1317-22.
3.   Heffner JE, Alberts WM, Irwin R, Wunderink R. Translating guidelines into clinical practice : recommendations to the American College of Chest Physicians. Chest 2000;118(2 Suppl):70S-3S.
4.   Jackson R, Feder G. Guidelines for clinical guidelines [editorial]. BMJ 1998;317:427-8.
5.   Saint S, Veenstra DL, Lipsky BA. The clinical and economic consequences of nosocomial central venous catheter-related infection: are antimicrobial catheters useful? Infect Control Hosp Epidemiol 2000;21:375-80.
6.   Mermel LA, Farr BM, Sherertz RJ, Raad II, O'Grady N, Harris JS, Craven DE. Guidelines for the management of intravascular catheter-related infections. Clin Infect Dis 2001;32:1249-72.
7.   O'Grady NP, Alexander M, Dellinger EP, Gerberding JL, Heard SO, Maki DG, Masur H, McCormick RD, Mermel LA, Pearson ML, et al. Guidelines for the prevention of intravascular catheter-related infections. Centers for Disease Control and Prevention. Morb Mortal Wkly Rep 2002;51(RR-10):1-29.
8.   Hugonnet S, Perneger TV, Pittet D. Alcohol-based handrub improves compliance with hand hygiene in intensive care units. Arch Intern Med 2002;162:1037-43.

9. Boyce JM, Pittet D. Guideline for Hand Hygiene in Health-Care Settings. Recommendations of the Healthcare Infection Control Practices Advisory Committee and the HICPAC/SHEA/APIC/IDSA Hand Hygiene Task Force. Society for Healthcare Epidemiology of America/Association for Professionals in Infection Control/Infectious Diseases Society of America. Morb Mortal Wkly Rep 2002;51(RR-16):1-45.
10. Sherertz RJ, Ely EW, Westbrook DM, Gledhill KS, Streed SA, Kiger B, Flynn L, Hayes S, Strong S, Cruz J, *et al.* Education of physicians-in-training can decrease the risk for vascular catheter infection. Ann Intern Med 2000;132:641-8.
11. Eggimann P, Harbarth S, Constantin MN, Touveneau S, Chevrolet JC, Pittet D. Impact of a prevention strategy targeted at vascular-access care on incidence of infections acquired in intensive care. Lancet 2000;355:1864-8.
12. Coopersmith CM, Rebmann TL, Zack JE, Ward MR, Corcoran RM, Schallom ME, Sona CS, Buchman TG, Boyle WA, Polish LB, *et al.* Effect of an education program on decreasing catheter-related bloodstream infections in the surgical intensive care unit. Crit Care Med 2002;30:59-64.
13. Raad I, Bodey GP. Infectious complications of indwelling vascular catheters. Clin Infect Dis 1992;15:197-210.
14. Garner JS, Jarvis WR, Emori TG, Toran TC, Hughes JM. CDC definitions for nosocomial infections. Am J Infect Control 1988;16:128-40.
15. Pittet D, Hugonnet S, Harbarth S, Mourouga P, Sauvan V, Touveneau S, Perneger TV, and the Members of the Infection Control Program. Effectiveness of a hospital-wide programme to improve compliance with hand hygiene. Lancet 2000;356:1307-12.
16. Veenstra DL, Saint S, Sullivan SD. Cost-effectiveness of antiseptic-impregnated central venous catheter for the prevention of catheter-related bloodstream infection. JAMA 1999;282:554-60.
17. Eggimann P, Hugonnet S, Harbarth S, Chraiti M.N, Touveneau S, Chevrolet JC, and Pittet D. Reduction of bloodstream infections 2 years following a global prevention strategy targeted at vascular access in ICU. Programs and Abstracts of the 41th Interscience Conference on Antimicrobial Agents and Chemotherapy, Chicago 2001. Abstract 2050
18. Greco PJ, Eisenberg JM. Changing physicians' practices. New Engl J Med 1993;329:1271-3.
19. Kretzer EK, Larson EL. Behavioral interventions to improve infection control practices. Am J Infect Control 1998;26:245-53.

Chapter 10

EDUCATION AS AN INTERVENTION FOR REDUCING VASCULAR CATHETER INFECTIONS

Robert J. Sherertz, M.D.
Infectious Diseases, Wake Forest University School of Medicine, Winston Salem, North Carolina

Introduction

Vascular catheter infections are important causes of patient morbidity and mortality which affect over 200,000 patients per year in the United States (1). A number of well done studies have identified interventions which reduce the risk of vascular catheter infections. These include the use of chlorhexidine as a skin preparation agent, the use of maximum sterile barriers for catheter insertion, and utilizing catheters with anti-infective coatings (2-7). With mounting evidence that various interventions are effective for reducing the risk of vascular catheter infections, guidelines have been developed that advise the medical community about the effectiveness of the various interventions and where they should be considered (8). One of the category I (strongly recommended) recommendations in the most recent guidelines from the CDC is that education about vascular catheter infections is an effective intervention that should be part of each hospital's infection control program (8). The remainder of this discussion will evaluate the available studies that have investigated whether education is an effective intervention for preventing vascular catheter infection.

Doing a literature review from 1966 to 2003 yielded eight studies where education was used in an attempt to reduce vascular catheter infections (Table) (9-16). Two of the investigations used education in response to extremely high infection rates (12,16). In the first study by Puntis et al. a 45% catheter infection rate in 58 patients was seen in children undergoing hyperalimentation therapy (16). An educational program was implemented by the nutritional care sister which included didactic instruction for all existing medical and nursing staff and was also given to all arriving new staff who participated in the insertion and care of vascular catheters. Education also included the use of a videotape about bag
changing and line flushing. Subsequent follow-up demonstrated a reduction in line sepsis to 8% in 107 patients. In the second study reported by Maas et al. a 42% infection rate (11/26) of central venous catheter-related bloodstream infection in a neonatal intensive care unit prompted an educational intervention (12). The infection control team delivered in-service training to both physicians and nurses including practical demonstrations of the more stringent catheter care protocol developed in response to the high infection rate. Over the subsequent four year follow-up period the infection rate was much lower (12%).

Table. Effect of Educational Interventions on Vascular Catheter Infection Rates.

Year	Educational Intervention	Group Educated	Patient Group	Study Length	#Patients Followed	Control Infect Rate[a]	Post Education Infection Rate[a]
2002(9)	Self-study module	RN	SICU	36m	4,283	10.8/1000dd	3.7/1000dd[c]
2000(10)	Didactic, hands-on	MD	ICU	30m	>7,000	4.5/1000pd	2.9/1000dd[c]
2000(11)	Didactic, hands-on	RN,MD	MICU	24m	3,154	11.3/1000pd	3.3/1000dd[c]
1998(12)	Didactic, hands-on	RN,MD	NICU	60m	182	42%(11/26)	12%/1000dd[c]
1997(13)	Didactic	RN,F	Ped	11m	286	4.6/1000dd	3.8/1000dd[d]
1996(14)	Didactic (RN) Hands-on (MS2)	RN,MD	All	72m	MA	0.9%	0.9%[d]
1994(15)	Didactic	RN	Mixed	6m	NA	0	0[d]
1991(16)	Didactic	RN,MD	Ped	NA	165	45%(26/58)	8%(9/107)[b,c]

Abbreviations: RN=registered nurse, MD=medical doctor, SICU=surgical intensive care unit, MICU=medical intensive care unit, Ped=pediatric, F=family, MS=medical student

a - Primary bloodstream infection rates; b - Catheter-related bloodstream infection rates; c - Significantly lower than control, P<.05; d - No difference from control, P>.05

In the other six investigations the educational interventions occurred in the absence of perceived excess infection rates and were implemented to see if the baseline infection rate could be reduced (9-11, 13-15). In three of the studies catheter-related bloodstream infection rates were not significantly reduced (13-15). Parras et al. evaluated the efficacy of an education program using two cross-sectional studies done before and after their intervention (15). The educational intervention consisted of didactic sessions by nurses given to all nurses working within a targeted 500 bed area of the hospital. No catheter-related bloodstream infection was documented either before or after the cross-sectional study. This study had very limited power to detect an impact on catheter-related bloodstream due to the cross-sectional design. Notably, they did document significant reductions in inappropriate catheter care (83% to 38%) and a reduction in skin colonization under the dressing (34% to 18%). No change was noted in catheter hub or catheter tip colonization. In the second study, Cohran et al. examined the impact of an intravascular surveillance and education program on catheter-related bloodstream infection (14). The educational component consisted of nurses giving feedback regarding compliance with standards of practice to physicians and nurses and didactic educational programs for physicians and nurses. There was no significant change in primary bloodstream infection rates comparing the three years during the program with three years after the program was discontinued for financial reasons. In the third study by Lange et al. an educational intervention was used to try and reduce the risk of catheter-related infections in a children's hospital (13). First a comprehensive assessment was made of what practices need to be changed. Then nurses educated nurses on eight nursing units as well as educated the families of all catheterized patients. The preintervention period was compared with the postintervention period and no difference was found in catheter-related bloodstream infection; although there was a significant reduction in catheter exit site infections (0.58 to 0.11).

In the final three studies educational interventions demonstrated reductions in catheter-related bloodstream infections from the baseline rate. Sherertz et al. developed a hands-on course whereby all arriving interns were taught the proper technique for doing a variety of basic procedures, one of which was the insertion of central lines (10). The course was initiated after it became clear that the use of maximal sterile barriers for central line insertion occurred with less than 20% of insertions in spite of yearly didactic education. The central line portion of the course was taught by intensive care physicians. Compared with prospectively collected baseline data the catheter-related bloodstream infection rate fell 35% during the 18 month follow-up period. They also documented a 48% increase in the use of full sterile drapes after the course began. Eggimann et al. developed a course taught to all medical intensive unit personnel targeting the reduction of vascular catheter infections (11). The course involved both didactic instruction and practical demonstrations. It was taught by a physician and two infection control nurses. In the targeted intensive care unit there was a significant reduction in primary bloodstream infection and catheter exit site infections, while no change occurred in a surgical intensive care unit where the intervention did not occur. Finally, Coopersmith et al. employed a self-study model that was given primarily to nurses working in a surgical intensive care unit (9). They found a 66% reduction in primary bloodstream infections comparing an 18 month baseline period with an 18 month post-intervention period. They observed no change in the rate of secondary bacteremias in the same time period.

In aggregate then 5/8 reports demonstrated a reduction in catheter-related infections and 3/8 found no change in infection rates. Of note 4/5 of the groups reporting success taught physicians with or without nurse education, whereas only 1/3 of the groups reporting no change taught physicians. This suggests that physician behavior plays a major role in the infection rates associated with ICU catheter-related infections. Since physicians are essentially involved only in inserting the catheters, this suggests that a successful ICU focused program must focus on the insertion process itself. This also suggests that most of the catheters in ICU patients are short term (<1-2 weeks), because there are good data available suggesting that after the first one to two weeks the predominant risk of infection is associated with line breaks which are primarily done by nurses (17).

That it was possible to change physician behavior through education with a resultant change in patient outcomes is a very important finding. There are many reports demonstrating the failure of physician CME in changing physician behavior and very few reports demonstrating changes in outcome (18-22). The recurring theme in the discussions of why CME is unsuccessful is that didactic interactions do not change physician practice. Programs that employ interactive or hands-on methodologies are much more likely to have successful outcomes (19). Thus, it may be significant that 4/5 successful programs reviewed in this article used more than didactic education and 2/3 unsuccessful programs used just didactic education. It is particularly interesting that the self-study modules employed by Coopersmith et al. and primarily used by nurses were able to have a significant effect in reducing the infection rates in a surgical ICU. It suggests that the nurses in the unit studied must play a very important role in the catheter insertion process, ie perhaps assisting the physician inserters, otherwise it is hard to imagine why there was such a large reduction in catheter-related infection.

The cost-benefit of a successful catheter infection education program is potentially large. Utilizing published attributable costs for catheter-related infections Sherertz et al. estimated that their program resulted in a net savings of at least $55,000 using the most conservative estimate of attributable cost with potential savings of up to $800,000 (10,22,23). Coopersmith et al. estimated that their intervention resulted in savings that could range from $185,000 to $2,808,000 (9). Thus, even though such educational programs may have significant start-up costs, the long term benefit potential can easily pay for the program with leftover additional savings.

In conclusion, studies in the literature demonstrate that educational programs can reduce the risk of catheter-related infections. Successful programs target physicians and use teaching methods in addition to didactic lectures, such as hands-on demonstrations. Successful programs are cost-effective and should be considered before jumping to more expensive interventions such as catheters with anti-infective coatings.

REFERENCES

1. Maki DG. Infections caused by intravascular devices used for infusion therapy. In: Waldvogel FA, Bisno AL, eds. Infections Associated with Indwelling Medical Devices. Washington, DC: ASM Press, 1994: 155-205.
2. Maki DG, Ringer M, Alvarado CJ. Prospective randomized trial of povidone-iodine, alcohol, and chlorhexidine for prevention of infection associated with central venous and arterial catheters. Lancet 1991;338:339-43.
3. Chaiyakunapruk N, Veenstra DL, Lipsky BA, Saint S. Chlorhexidine compared with povidone-iodine solution for vascular catheter-site care: a meta-analysis. Ann Intern Med 2002;136:792-801.
4. Raad II, Hohn DC, Gilbreath BJ, et al. Prevention of central venous catheter-related infections by using maximal sterile barrier precautions during insertion. Infect Control Hosp Epidemiol 1994;15:231-8.

5. Maki DG, Stolz SM, Wheeler S, Mermel LA. Prevention of central venous catheter-related bloodstream infection by use of an antiseptic-impregnated catheter. A randomized, controlled trial. Ann Intern Med 1997;127:257-66.
6. Raad I, Darouiche R, Dupuis J, et al. Central venous catheters coated with minocycline and rifampin for the prevention of catheter-related colonization and bloodstream infections. A randomized, double-blind trial. The Texas Medical Center Catheter Study Group. Ann Intern Med 1997;127:267-74.
7. Darouiche RO, Raad II, Heard SO, et al. A comparison of two antimicrobial-impregnated central venous catheters. Catheter Study Group. N Engl J Med 1999;340:1-8.
8. O'Grady NP, Alexander M, Dellinger EP, et al. Guidelines for the prevention of intravascular catheter-related infections. MMWR Morb Mortal Wkly Rep 2002;51(RR-10):1-29. Available online at: http://www.cdc.gov/ncidod/hip/iv/iv.htm.
9. Coopersmith CM, Rebmann TL, Zack JE, et al Effect of an education program on decreasing catheter-related bloodstream infections in the surgical intensive care unit. Crit Care Med 2002;30:59-64.
10. Sherertz RJ, Ely EW, Westbrook DM, et al. Education of physicians-in-training can decrease the risk for vascular catheter infection. Ann Intern Med 2000;132:641-8.
11. Eggimann P, Harbarth S, Constantin M-N, Touveneau S, Chevrolet J-C, Pittet D. Impact of a prevention strategy targeted at vascular-access care on incidence of infections acquired in intensive care. Lancet 2000;355:1864-8.
12. Maas A, Flament P, Pardou A, Deplano A, Dramaix M, Struelens MJ. Central venous catheter-related bacteraemia in critically ill neonates: risk factors and impact of a prevention programme. J Hosp Infect 1998;40:211-24.
13. Lange BJ, Weiman M, Feuer EJ, et al. Impact of changes in catheter management on infectious complications among children with central venous catheters. Infect Control Hosp Epidemiol 1997;18:326-32.
14. Cohran J, Larson E, Roach H, Blane C, Pierce P. Effect of intravascular surveillance and education program on rates of nosocomial bloodstream infections. Issues in Infect Control 1996;25:161-4.
15. Parras F, Ena J, Bouza E, et al. Impact of an educational program for the prevention of colonization of intravascular catheters. Infect Control Hosp Epidemiol 1994;15:239-42.
16. Puntis JWL, Holden CE, Smallman S, Finkel Y, George RH, Booth IW. Staff training: a key factor in reducing intravascular catheter sepsis. Arch Dis Child. 1991;66:335-7.
17. Sherertz RJ. Pathogenesis of vascular catheter-related infections. Ed. by Seifert H, Jansen B, Farr BM. In Catheter-Related Infections. Marcel Dekker, Inc., New York, 1997, pp1-30.
18. Davis DA, Thomson MA, Oxman AD, Haynes RB. Changing physician performance. A systematic review of the effect of continuing medical education strategies. JAMA. 1995 Sep 6;274(9):700-5.
19. Davis D, O'Brien MAT, Freemantle N, Wolf FM, Mazmanian P, Taylor-Vaisey A. Impact of formal continuing medical education. Do conferences, workshops, rounds, and other traditional continuing education activities change physician behavior or health care outcomes. JAMA 1999;282:867-74.

20.   Thomson O'Brien MA, Freemantle N, Oxman AD, Wolf F, Davis DA, Herrin J. Continuing education meetings and workshops: effects on professional practice and health care outcomes. Cochrane Database Syst Rev. 2001;(2):CD003030.

21.   Tu K, Davis. Can we alter physician behavior by educational methods? Lessons learned from studies of the management and follow-up of hypertension. J Contin Educ Health Prof. 2002 Winter;22(1):11-22.

22.   Cauffman JG, Forsyth RA, Clark VA, Foster JP, Martin KJ, Lapsys FX, Davis DA. Randomized controlled trials of continuing medical education: what makes them most effective? J Contin Educ Health Prof. 2002 Fall;22(4):214-21.

23.   Haley RW, Culver DH, White JW, et al. The efficacy of infection surveillance and control programs in preventing nosocomial infections in US hospitals. Am J Epidemiol. 1985;121:182-205.

24.   Pittet D, Wenzel RP. Nosocomial bloodstream infections in the critically ill. JAMA 1994;272:1820.

Chapter 11

ICU PREVENTION STRATEGIES

Jean-François Timsit, Ph.D.
Réanimation Médicale et Infectieuse, Hôpital Bichat-Claude Bernard
Paris, France

Introduction

Central venous catheters (CVCs) inserted for short-term use are common and indispensable tools for the care of critically ill patients. The risk of exposure to these devices is 48% per intensive care unit (ICU) day. However, short-term CVCs are also associated with serious complications, notably infection, and catheter-related bloodstream infection (CR-BSI) represents the third most frequent cause of nosocomial infection in the ICU.

Risk factors for CVC-BSI can be divided into host-related and catheter-related factors. The underlying condition of the patient influences the risk of CR-BSI. Malignancy, immunodeficiency, severe burns, and malnutrition, all conditions that compromise host defenses, lead to a higher rate of infection. Similarly, severe sepsis, or severe and sustained multiple organ dysfunction are also associated with a higher risk of CR-BSI (1).

A large number of risk factors are device-related, thus suggesting that CR-BSI is accessible to prevention if rigorous policies are adopted. Healthcare worker education and training, and continuous quality improvement programs are essential and discussed in detail in another chapter. The potential interest of different type of procedures and technical developments usable in ICU patients have been extensively reviewed (1-3). In this chapter, some particular points will be looked at in detail, together with practical recommendations for implementing and encouraging an ICU prevention strategy.

MECHANISMS OF INFECTION

Colonization of the catheter may occur by two main pathways: the extraluminal route or the intraluminal route. For short-term CVSs (<15-20 days), the cutaneous entry site is the predominant route of colonization, whereas colonization via the endoluminal route resulting from hub contamination predominates for long-term CVCs (4). In both cases, the source of microorganisms is from the patient's own skin commensal flora. Accordingly, the occurrence of bacteremia caused by common skin organisms is a major criterion for the diagnosis of CR-BSI. Most infections are caused by Gram-positive cocci.

Other sources of microorganisms, which are considered to be relatively minor contributors to sepsis, are either via hematogeneous spread from other body sites or from a contaminated infusate.

CATHETER INSERTION

Sterile Barrier Precautions

Full barrier precautions using sterile gloves, long-sleeved sterile gown, mask, cap and large sterile sheet drape during catheter insertion is a fundamental way to prevent CR-BSI and must be considered as standard during central venous and pulmonary catheter insertion (5). The potential interest of handrubbing with an aqueous alcoholic solution in improving the compliance and the tolerance of hand-cleansing in the operating room have been reported recently, and its use might probably be extended to central venous catheter insertion precautions (6).

Cutaneous Antisepsis

Rigorous cleansing and disinfection of the insertion site is a key point. A recent meta-analysis suggested that chlorhexidine solution is superior to povidone iodine solution. However, most studies compared 2% chlorhexidine or chlorhexidine-alcohol preparation to povidone iodine (7). Alcoholic-povidone iodine is associated with a better cutaneous antisepsis (8). and a lower rate of significant catheter colonization than the regular solution of povidone iodine, (9) but the former has not been compared as yet to chlorhexidine-alcohol.

In current practice, chlorhexidine solutions are recommended. However, povidone iodine is used in many ICUs because of its broad spectrum activity and a very rare intrinsic or acquired resistance to bacteria or fungi (10). If used, povidone iodine needs to remain on the skin for at least 1 minute before catheter insertion. The interest of alcoholic-povidone iodine solution needs to be tested in clinical trials to determine its actual efficacy.

Catheter Insertion Site

Central venous catheter insertion is required in many critically ill patients. Selection of the insertion site should be based on both the ease and the risks of the procedure. These latter include infection, thrombosis, and mechanical complications (11).

The choice of the best central venous access for a particular patient is based on the rate and the severity of failures and complications. Internal jugular access is associated with a low rate of severe mechanical complications in the ICU as compared to subclavian access and should be preferred for short-term access (<5-7 days) and for hemodialysis catheters (12). Subclavian access is associated with a lower risk of infection and is the route of choice in experienced hands if the risk of infection is high (CVC > 5-7 days), or if the risk of mechanical complication is low.1 A recent prospective, randomized study in 289 adult ICU patients compared the untunnelled subclavian approach to the untunnelled femoral approach (13). Patients with severe hypoxia (PF ratio<150 mmHg) or coagulation disorders (platelets<50 000/mm^3, PTT>1.6 normal, APTT>2 normal, anticoagulant therapy) were not included. The femoral approach was associated with higher rates of significant catheter colonization (19.8% vs. 4.5%, p<0.001) and CR-BSI (4.4% vs. 1.5%, p=0.07). In the same study, an independent positive association was found between catheter-related thrombosis and the femoral approach (21.5% vs. 1.9%, p<0.001), and complete thrombosis was diagnosed in 6% of patients in the femoral group vs. none in the subclavian group (p=0.01). Finally, the risk of major mechanical complications was not significantly different between the groups (subclavian, 4/144 (4 pneumothoraces) vs. femoral, 2/145 (2 hematomas requiring blood transfusion and/or surgery), p=0.45). In mechanically ventilated patients without severe coagulation disorders or respiratory failure, subclavian access should be preferred over femoral access. The use of femoral catheters should be restricted to patients for whom pneumothorax or hemorrhage would be unacceptable.

Catheter Tunneling

Subcutaneous tunnelling of short-term catheters inserted into the jugular (14) or the femoral vein (15) decreased the risk of CR-BSI more than 3-fold . It probably acts by reducing the extraluminal migration of microorganisms into the bloodstream and by allowing a better occlusion of dressings. The interest of catheter tunnelling is maximized if the main route of catheter contamination is extraluminal and when catheters are not used for drawing blood (1). It should be recommended for short-term CVCs (<30 days) inserted in the ICU if subclavian access is undesirable.

PROPHYLACTIC TREATMENTS

Antithrombotic Prophylaxis

Both experimental and cohort studies (16,17) have suggested a close relationship between catheter-thrombosis and infection. Several proteins of the thrombus are able to increase adherence of staphylococci and Candida spp. to catheters and thrombus formation on indwelling intravascular catheters is associated with CR-BSI.

In a meta-analysis, prophylactic heparin (3 U/ml in parenteral nutrition, 5000 U q 6 or 12 hours flush, or 2500 U low molecular weight heparin subcutaneously or heparin-bonded catheters) was shown to reduce catheter-related thrombosis (RR=0.4 [95%CI: 0.2-0.8], catheter colonization (RR=0.6 [95%CI: 0.06-0.6]), and although insignificantly, CR-BSI (RR=0.26 [95% CI: 0.07-1.03]) (18) Subcutaneous, low-molecular-weight heparins appeared to be as effective as non-fractioned heparins. However, the studies included in this meta-analysis are old and the results of a more recent report (19) could be discussed (more internal jugular catheters in the control group and an analysis performed on 33/55 included patients). Since most heparin solutions contain preservatives with antimicrobial activity, it is unclear if any decrease in the rate of CR-BSI would be due to the reduced thrombus formation, the preservative, or both.

However, more recently, a randomized double blind study in critically ill children showed that heparin-bonded catheters reduced thrombosis (0 vs 8%, p=0.006) and the frequency of positive blood cultures (drawn through the catheter) (4 vs 33%, p<0.0005) (20). These encouraging results need to be confirmed in adult ICU patients.

CATHETER CARE

Catheter Replacement

Although not accessible to preventive measures, repeated catheterization increases the risk of catheter infection. This finding argues, together with randomized study, to avoid any routine replacement of CVCs (21) for catheters that are functioning and have no evidence of local or systemic complications. Physicians and nurses should assess patients' need for intravascular catheters on a daily basis.

Catheter Dressing and Tubing

Semi-permeable transparent dressings are widely used and allow continuous observation of the skin insertion site and reduce the risk of extrinsic colonization. However, the first generation transparent dressings promoted moisture and bacterial proliferation and have been associated with a higher rate of infection when compared with traditional gauze dressings (22). However, Maki *et al.*, using highly semi-permeable transparent dressings did not find any differences with gauze dressings (23). The choice of catheter dressings might be a matter of preference. Gauze dressing is usually preferred if blood is oozing from the catheter insertion site. Catheter dressings should be changed immediately if the dressing becomes damp, loosened or soiled. The optimal frequency of routine changes of CVC dressing is not known but it should be at least every 48-72 hours.

Antibiotic and iodophor ointments are not useful and could increase the risk of infection by *Candida* sp. (2,3,24).

In a randomized trial in adult ICU patients, the chlorhexidine-impregnated sponge dressing resulted in a 60% reduction in CR-BSIs

152 *Catheter-Related Infections in the Critically Ill*

compared to standard non-medicated polyurethane dressings (25). In another study, neonates randomized to the antimicrobial dressing group were less likely to have colonized CVC tips than control neonates (15.0% vs 24.0%, relative risk [RR]: 0.6 95% confidence interval [CI]: 0.5-0.9), but rates of CR-BSI (3.8% vs 3.2%, RR: 1.2, CI: 0.5-2.7) and BSI without an identified source (15.2% vs 14.3%, RR: 1.1, CI: 0.8-1.5) were unchanged. Moreover, localized contact dermatitis from the antimicrobial dressing, requiring crossover into the control group, occurred in 15 (15.3%) of 98 exposed neonates weighing less than or equal to 1000 g. Differences in catheter tip colonization rates were more obvious in catheters in situ for less than 14 days (26). It is recommended that tubings be replaced at least every 72 hours. However, tubing changes delayed to every 4 days (instead of 2 days) did not increase the rate of CR-BSI.4 Nevertheless, tubings used to administer blood, blood products or lipid emulsions (including propofol infusions) should be replaced within 24 hours.

A 4-fold reduction in the rate of CR-BSI was reported in a cohort study with the use of a new antiseptic hub model (containing an antiseptic chamber filled with 3% iodinated alcohol) in a surgical unit (27). On the other hand, a recent clinical trial failed to show any benefit from the use of this hub in preventing CR-BSI.28 The potential interest of this new hub model is probably more important for long-term CVCs outside the ICU.

Overall, any excessive manipulation of CVCs independently increases the risk for CR-BSI and must be avoided (24). The effect of a hands-off system on manipulation have been tested for pulmonary artery catheters. The use of this system decreased dramatically the rate of PAC-BSI (29).

Collagen-Silver Impregnated Cuffs

A silver impregnated collagen cuff attached to CVCs and left below the skin insertion site significantly decreased the risk of extraluminal colonization associated with short-term catheters (mean duration of placement <10 days). The ionic silver has broad spectrum activity against bacteria and fungi, and the cuff provides a mechanical barrier to the migration of microorganisms along the external surface of the catheter. The collagen is biodegradable in nature and the silver ions chelated to the cuff are released completely within 3 to 7 days (30-33).

Extrusion of the collagen cuff after insertion is frequent when inserted by an inexperienced physician. In two randomized clinical trials conducted in surgical patients assigned to receive a CVC with or without a silver cuff, the incidence of CR-BSI was significantly greater in the control versus the cuffed catheter group (30,31). On the other hand, no difference in the rates of catheter colonization or incidence of CR-BSI was observed between patients who received and those who did not receive a silver cuffed CVC (32).

Antimicrobial-Coated or –Impregnated Catheters

The efficacy of catheters impregnated with chlorhexidine and silver-sulfadiazine on the external surface has been tested in many randomized studies. In experimental studies, a steep exponential decrease in antimicrobial activity was observed during the first week of insertion for devices treated with chlorhexidine-silver-sulfadiazine. A well-conducted meta-analysis concluded that this technique reduces the risk of CR-BSI (RR=0.4, 95%CI: 0.2-0.8) for short-term central venous catheterisation (34). A more recent meta-analysis showed that the risk of BSI decreased only in trials with the shorter average insertion time (median 6 days: 5.2 to 7.5; from 4.1 to 1.9%; OR: 0.48 95% CI: 0.25-0.91) as compared to others with a longer duration (median 12 days: 7.8 to 20 days; from 4.5% to 4.2%; OR: 0.94 95% CI; 0.58-1.54) (35). This technique is cost-saving in settings where the incidence of CR-BSI of short-term CVCs is high (more than 3.3 episodes per 1000 catheter-days) and when the average insertion time is less than 8 days. New chlorhexidine-sulfadiazine-impregnated catheters are now developed with an impregnation at both the internal and external surfaces. Results presented during the 41st and the 42nd ICAAC meetings suggested that these catheters are effective even when the average insertion time is 12 to 15 days.

Resistance to chlorhexidine-sulfadiazine has not been demonstrated in clinical studies. However, resistance to chlorhexidine has been induced in *in vitro* studies (36). Rare cases of anaphylactic reaction to the chlorhexidine component of this catheter has been reported (24). Consequently, the use of chlorhexidine-silver-sulfadiazine impregnated on the external surface catheters should be used when catheter is expected to last less than 8 days and when the rate of infection is high, despite adherence to other strategies such as maximal barrier precautions and implementation of an educational program. As acceptable incidences are, at least in France, between 1 and 3

CR-BSI per 1000 catheter-days, (3) the use of such impregnated catheters are not the standard.

Catheters impregnated intraluminally and extraluminally with minocycline-rifampin reduced the risk or CR-BSI as compared to polyurethane catheters and to externally coated chlorhexidine-silver-sulfadiazine-impregnated catheters (0.3% vs 3.4%, p<0.002) (37,38). The superiority of antimicrobial-impregnated catheters was probably due to the absence of intraluminal impregnation with chlorhexidine-sulfadiazine. The impact of antimicrobial-impregnated CVCs on resistance on skin flora is not sufficiently known to recommend its use.

How to Implement Your Own ICU Prevention Strategy?

Three general preventive measures are assumed when implementing practical recommendations for catheter insertion and surveillance: prevention and control of multiresistant bacteria spread, hand hygiene and surveillance of nosocomial infections. These seemingly trivial and general measures are essential and are extensively detailed elsewhere. If unobserved, any attempts in implementing catheter-specific prevention strategy will be futile (39).

Healthcare workers should first focus on the several established methods primarily directed at preventing extraluminal contamination of the catheter, which predominates for short-term catheters. Compliance with these measures should allow satisfactory control of the rate of CR-BSI in ICU patients. Strong educational efforts to obtain the compliance of each care provider to protocols must be regularly discussed and updated, and continuous surveillance of CVC infection rates and feedback should be instituted (Table).

As understaffing has been repeatedly identified as a risk factor for CR-BSI (40,41) the organisational aspect of the ICU should be considered.

If the impact of such measures remain unsatisfactory, more specific techniques (such as the use of chlorhexidine-silver-sulfadiazine-coated catheters) might be considered.

CONCLUSIONS

CR-BSI remains a leading cause of nosocomial infections, particularly in ICUs, and is the most frequent cause of hospital-acquired bacteremia. Although CR-BSI have milder consequences than other bacteremic infections, it is a typical device-associated iatrogenic infection and therefore mostly accessible to prevention, if rigorous policies are adopted. It should be one of the main targets of quality improvement programs.

REFERENCES

1. Mermel LA. Prevention of intravascular catheter-related infections. Ann Intern Med 2000; 132: 391-402.
2. O'Grady NP, Alexander M, Dellinger EP, Gerberding JL, Heard SO, Maki DG, Masur H, McCormick RD, Mermel LA, Pearson ML, Raad, II, Randolph A and Weinstein RA. Guidelines for the prevention of intravascular catheter-related infections. Centers for Disease Control and Prevention. MMWR Recomm Rep 2002; 51: 1-29.
3. Timsit JF. Réactualisation de la XIIe Conférence de Consensus de la S.R.L.F.. Infections liées aux cathéters veineux centraux en réanimation. Réanimation 2003; 12: in press.
4. Sitges-Serra A, Linares J, Perez JL, Jaurrieta E and Lorente L. A randomized trial on the effect of tubing changes on hub contamination and catheter sepsis during parenteral nutrition. JPEN J Parenter Enteral Nutr 1985; 9: 322-5.
5. Raad, II, Hohn DC, Gilbreath BJ, Suleiman N, Hill LA, Bruso PA, Marts K, Mansfield PF and Bodey GP. Prevention of central venous catheter-related infections by using maximal sterile barrier precautions during insertion. Infect Control Hosp Epidemiol 1994; 15: 231-8.
6. Parienti JJ, Thibon P, Heller R, Le Roux Y, von Theobald P, Bensadoun H, Bouvet A, Lemarchand F and Le Coutour X. Hand-rubbing with an aqueous alcoholic solution vs traditional surgical hand-scrubbing and 30-day surgical site infection rates: a randomized equivalence study. JAMA 2002; 288: 722-7.
7. Chaiyakunapruk N, Veenstra DL, Lipsky BA and Saint S. Chlorhexidine compared with povidone-iodine solution for vascular catheter-site care: a meta-analysis. Ann Intern Med 2002; 136: 792-801.
8. Calfee DP and Farr BM. Comparison of four antiseptic preparations for skin in the prevention of contamination of percutaneously drawn blood cultures: a randomized trial. J Clin Microbiol 2002; 40: 1660-5.
9. Parienti JJ, Ducheyron D., M. R, Malbruny B, Leclercq R, Lecontour X and Charbonneau P. Prospective Randomized Trial of 10% Povidone-Iodine (PVI) versus

5% Alcohol Plus 5% Povidone-Iodine (APVI) for Prevention of Central Venous Catheter (CVC) Colonization.2002;318.

10. McDonnell G and Russell AD. Antiseptics and disinfectants: activity, action, and resistance. Clin Microbiol Rev 1999; 12: 147-79.

11. Timsit JF. Central venous access in intensive care unit patients: is the subclavian vein the royal route? Intensive Care Med 2002; 28: 1006-8.

12. Ruesch S, Walder B and Tramer MR. Complications of central venous catheters: internal jugular versus subclavian access--a systematic review. Crit Care Med 2002; 30: 454-60.

13. Merrer J, De Jonghe B, Golliot F, Lefrant JY, Raffy B, Barre E, Rigaud JP, Casciani D, Misset B, Bosquet C, Outin H, Brun-Buisson C and Nitenberg G. Complications of femoral and subclavian venous catheterization in critically ill patients: a randomized controlled trial. JAMA 2001; 286: 700-7.

14. Timsit JF, Sebille V, Farkas JC, Misset B, Martin JB, Chevret S and Carlet J. Effect of subcutaneous tunneling on internal jugular catheter-related sepsis in critically ill patients: a prospective randomized multicenter study. JAMA 1996; 276: 1416-20.

15. Timsit JF, Bruneel F, Cheval C, Mamzer MF, Garrouste-Orgeas M, Wolff M, Misset B, Chevret S, Regnier B and Carlet J. Use of tunneled femoral catheters to prevent catheter-related infection. A randomized, controlled trial. Ann Intern Med 1999; 130: 729-35.

16. Raad, II, Luna M, Khalil SA, Costerton JW, Lam C and Bodey GP. The relationship between the thrombotic and infectious complications of central venous catheters. Jama 1994; 271: 1014-6.

17. Timsit JF, Farkas JC, Boyer JM, Martin JB, Misset B, Renaud B and Carlet J. Central vein catheter-related thrombosis in intensive care patients: incidence, risks factors, and relationship with catheter-related sepsis. Chest 1998; 114: 207-13.

18. Randolph AG, Cook DJ, Gonzales CA and Andrew M. Benefit of heparin in central venous and pulmonary artery catheters: a meta-analysis of randomized controlled trials. Chest 1998; 113: 165-71.

19. Appelgren P, Ransjo U, Bindslev L, Espersen F and Larm O. Surface heparinization of central venous catheters reduces microbial colonization *in vitro* and *in vivo*: results from a prospective, randomized trial. Crit Care Med 1996; 24: 1482-9.

20. Pierce CM, Wade A and Mok Q. Heparin-bonded central venous lines reduce thrombotic and infective complications in critically ill children. Intensive Care Med 2000; 26: 967-72.

21. Timsit JF. Scheduled replacement of central venous catheters is not necessary. Infect Control Hosp Epidemiol 2000; 21: 371-4.

22. Hoffmann KK, Weber DJ, Samsa GP and Rutala WA. Transparent polyurethane film as an intravenous catheter dressing. A meta-analysis of the infection risks. JAMA 1992; 267: 2072-6.

23. Maki DG, Stolz SS, Wheeler S and Mermel LA. A prospective, randomized trial of gauze and two polyurethane dressings for site care of pulmonary artery catheters: implications for catheter management. Crit Care Med 1994; 22: 1729-37.

24. Mermel LA, Farr BM, Sherertz RJ, Raad, II, O'Grady N, Harris JS and Craven DE. Guidelines for the management of intravascular catheter-related infections. Infect Control Hosp Epidemiol 2001; 22: 222-42.

25. Maki D, LA. M, Kluger D, Narins L, Knasinski V, Parentau S and Covington P.The efficacy of a chlorexidine-impregnated sponge (Biopatch) for the prevention of intravascular catheter-related infection: a prospective, randomized, controlled, multicenter study [abstract].2000.

26. Garland JS, Alex CP, Mueller CD, Otten D, Shivpuri C, Harris MC, Naples M, Pellegrini J, Buck RK, McAuliffe TL, Goldmann DA and Maki DG. A randomized trial comparing povidone-iodine to a chlorhexidine gluconate-impregnated dressing for prevention of central venous catheter infections in neonates. Pediatrics 2001; 107: 1431-6.

27. Segura M, Alvarez-Lerma F, Tellado JM, Jimenez-Ferreres J, Oms L, Rello J, Baro T, Sanchez R, Morera A, Mariscal D, Marrugat J and Sitges-Serra A. A clinical trial on the prevention of catheter-related sepsis using a new hub model. Ann Surg 1996; 223: 363-9.

28. Luna J, Masdeu G, Perez M, Claramonte R, Forcadell I, Barrachina F and Panisello M. Clinical trial evaluating a new hub device designed to prevent catheter-related sepsis. Eur J Clin Microbiol Infect Dis 2000; 19: 655-62.

29. Cohen Y, Fosse JP, Karoubi P, Reboul-Marty J, Dreyfuss D, Hoang P and Cupa M. The "hands-off" catheter and the prevention of systemic infections associated with pulmonary artery catheter: a prospective study. Am J Respir Crit Care Med 1998; 157: 284-7.

30. Flowers RH, 3rd, Schwenzer KJ, Kopel RF, Fisch MJ, Tucker SI and Farr BM. Efficacy of an attachable subcutaneous cuff for the prevention of intravascular catheter-related infection. A randomized, controlled trial. JAMA 1989; 261: 878-83.

31. Maki DG, Cobb L, Garman JK, Shapiro JM, Ringer M and Helgerson RB. An attachable silver-impregnated cuff for prevention of infection with central venous catheters: a prospective randomized multicenter trial. Am J Med 1988; 85: 307-14.

32. Smith HO, DeVictoria CL, Garfinkel D, Anderson P, Goldberg GL, Soeiro R, Elia G and Runowicz CD. A prospective randomized comparison of an attached silver-impregnated cuff to prevent central venous catheter-associated infection. Gynecol Oncol 1995; 58: 92-100.

33. Bonawitz SC, Hammell EJ and Kirkpatrick JR. Prevention of central venous catheter sepsis: a prospective randomized trial. Am Surg 1991; 57: 618-23.

34. Veenstra DL, Saint S, Saha S, Lumley T and Sullivan SD. Efficacy of antiseptic-impregnated central venous catheters in preventing catheter-related bloodstream infection: a meta-analysis. JAMA 1999; 281: 261-7.

35. Walder B, Pittet D and Tramer MR. Prevention of bloodstream infections with central venous catheters treated with anti-infective agents depends on catheter type and insertion time: evidence from a meta-analysis. Infect Control Hosp Epidemiol 2002; 23: 748-56.

36. Tattawasart U, Maillard JY, Furr JR and Russell AD. Development of resistance to chlorhexidine diacetate and cetylpyridinium chloride in Pseudomonas stutzeri and changes in antibiotic susceptibility. J Hosp Infect 1999; 42: 219-29.

37.   Darouiche RO, Raad, II, Heard SO, Thornby JI, Wenker OC, Gabrielli A, Berg J, Khardori N, Hanna H, Hachem R, Harris RL and Mayhall G. A comparison of two antimicrobial-impregnated central venous catheters. Catheter Study Group. N Engl J Med 1999; 340: 1-8.

38.   Raad I, Darouiche R, Dupuis J, Abi-Said D, Gabrielli A, Hachem R, Wall M, Harris R, Jones J, Buzaid A, Robertson C, Shenaq S, Curling P, Burke T and Ericsson C. Central venous catheters coated with minocycline and rifampin for the prevention of catheter-related colonization and bloodstream infections. A randomized, double-blind trial. The Texas Medical Center Catheter Study Group. Ann Intern Med 1997; 127: 267-74.

39.   Eggimann P, Harbarth S, Constantin MN, Touveneau S, Chevrolet JC and Pittet D. Impact of a prevention strategy targeted at vascular-access care on incidence of infections acquired in intensive care. Lancet 2000; 355: 1864-8.

40.   Robert J, Fridkin SK, Blumberg HM, Anderson B, White N, Ray SM, Chan J and Jarvis WR. The influence of the composition of the nursing staff on primary bloodstream infection rates in a surgical intensive care unit. Infect Control Hosp Epidemiol 2000; 21: 12-7.

41.   Fridkin SK, Pear SM, Williamson TH, Galgiani JN and Jarvis WR. The role of understaffing in central venous catheter-associated bloodstream infections. Infect Control Hosp Epidemiol 1996; 17: 150-8.

Chapter 12

NOVEL STRATEGIES OF PREVENTING CATHETER-RELATED INFECTIONS IN THE ICU

Naomi P. O'Grady, M.D.
Warren Magnusen Clinical Center, Critical Care Medicine Department, National Institutes of Health, Bethesda, Maryland

Introduction

Central venous catheters are essential in medical practice, particularly in intensive care units (ICUs). While such catheters provide necessary vascular access, the use of these catheters puts patients at risk for a variety of complications, including local site infection, catheter-related bloodstream infection (CR-BSI), septic thrombophlebitis, endocarditis, and metastatic infection (e.g., lung abscess, brain abscess, osteomyelitis, endophthalmitis). As healthcare-acquired infection rates become more frequently included as benchmarks for assessing the quality of patient care, implementation of proven prevention strategies takes on an added importance.

Data from the Surveillance and Control of Pathogens of Epidemiologic Importance (SCOPE) show that 70 percent of healthcare-acquired bloodstream infections occur in patients with central venous catheters (1). Associated with each infection is an average cost of between $34,508-$56,000 and an addition of 7 days to the hospitalization (2-4). Reducing the risk of catheter-related infection should be part of a comprehensive infection control program to reduce the incidence of health-care acquired infection and antimicrobial resistance in the ICU.

In the ICU setting the incidence of infection is often higher than in the less acute in-patient or in ambulatory settings. Central venous catheters are estimated to be present in 50% of all ICU patients. Rates of catheter-associated bloodstream infections range from 2.9 (cardiothoracic ICU) to 11.3 (high-risk nursery, <1000 grams) per 1000 central catheter days (5), with the incidence varying considerably with the type of catheter, frequency of catheter manipulation, and patient-related factors such as underlying disease and acuity of illness. In addition to patients being sicker in the ICU setting, central venous access may be needed for extended periods of time, patients are often colonized with healthcare-acquired organisms, and catheters may be manipulated multiple times per day for the administration of fluids, drugs, and blood products. Moreover, some catheters may be inserted in urgent situations during which optimal attention to aseptic technique may not be feasible, augmenting the potential for contamination and subsequent clinical infection. Thus the ICU provides a ripe setting for catheter-related infections.

EPIDEMIOLOGY AND MICROBIOLOGY

The most common organisms causing nosocomial healthcare-acquired BSIs have been changing over the last 15 years. Between 1986 and 1989, coagulase-negative staphylococci followed by S. aureus were the most frequently reported causes of BSIs. Pooled data from 1992- 1999 indicate that coagulase-negative staphylococci followed by enterococci are now the most frequently isolated causes of nosocomial BSIs. Coagulase-negative staphylococci now accounts for 37 % while S. aureus now accounts for only 12.6% of reported nosocomial BSIs (6). Also notable was the susceptibility pattern of the *S. aureus* isolates. In 1999, for the first time since NNIS has been reporting susceptibilities, more than 50% of all *S. aureus* nosocomial ICU isolates in the United States were oxacillin resistant (6).

Enterococci, the second most commonly reported bloodstream pathogen in the NNIS database from 1992-1999 accounted for 13.5% of BSIs, an increase from 8% reported to NNIS from 1986 through 1989. There has also been a dramatic rise in enterococcal ICU isolates resistant to vancomycin, escalating from 0.5% in 1989 to 25.9% in 1999 (6).

The frequency of fungal pathogens causing BSI have been stable over the last 15 years, with Candida making up 8% of nosocomial BSIs over this period of time. However, resistance of Candida species to commonly used antifungal agents has become a clinically important issue. *C.albicans* bloodstream isolates from hospitalized patients are becoming resistant to fluconazole as determined by in vitro susceptibility testing. Additionally, almost 50% of Candida BSIs are caused by non-albicans species including *C. glabrata* and *C. krusei*, which are more likely than *C. albicans* to demonstrate resistance to fluconazole and itraconazole.

Gram negative bacilli accounted for 19% of BSIs between 1986 and 1989 (7), compared to 14% of BSIs between 1992-1999 (6). An increasing fraction of ICU related isolates are caused by enterobacteriaceae -producing extended-spectrum b-lactamases (ESBLs), particularly *K. pneumoniae* . Such organisms are resistant to many commonly used cephalosporins.

PATHOGENESIS

Migration of skin organisms at the insertion site into the cutaneous catheter tract with colonization of the catheter tip is the most common route of infection for peripherally inserted, short-term catheters (8,9). Contamination of the catheter hub is an important contributor to intraluminal colonization of long-term catheters (10). Occasionally, catheters may become hematogenously seeded from another focus of infection. Rarely, infusate contamination leads to CR-BSI.

Important pathogenic determinants of catheter-related infection are (1) the material of which the device is made and (2) the intrinsic properties of the infecting organism. Some catheter materials also have irregularities on the catheter surface that enhance the microbial adherence of certain species (e.g., coagulase-negative staphylococci, *Acinetobacter calcoaceticus*, and *Pseudomonas aeruginosa)* (11-13). Catheters made of certain materials may be especially vulnerable to microbial colonization and subsequent infection. Additionally, some catheter materials are more thrombogenic than others, a characteristic that also may predispose to catheter colonization and catheter-related infection (13,14). This association has led to emphasis on preventing catheter-related thrombus as a mechanism for reducing CR-BSI, although there is little clinical data that supports this practice.

NOVEL STRATEGIES FOR PREVENTION OF CATHETER-RELATED INFECTIONS

Continuing Education and Quality Assurance

Although not particularly novel, the role of continuing education cannot be overemphasized in preventing catheter-related infections. As institutions begin to use infection rates for benchmarking quality of care and patient safety, applying simple concepts of infection control strategies takes on added significance. As knowledge, technology and health care settings change, infection control and prevention measures must change. This implies the need for well-organized programs that provide, monitor and evaluate care, and to provide education to all caregivers. Data collected over the last 15 years have consistently found that risk of infection declines following standardization of aseptic care and that insertion and maintenance of intravascular catheters by educated and trained staff may decrease the risk of catheter colonization and CR-BSI (15-18). Additionally, specialized "IV teams" have shown unequivocal effectiveness in reducing the incidence of catheter-related infections and associated complications and costs (19-21). Health care workers should be educated regarding indications for intravascular catheter use, proper procedures for the insertion and maintenance of intravascular catheters and appropriate infection control measures to prevent intravascular catheter-related infections..

Bedside Ultrasound

The availability of bedside ultrasound may play a role in the site selection for central venous catheter placement, thus avoiding unnecessary attempts at sites not suitable for catheter placement, and likewise avoiding fewer attempts at sites that are suitable. Although there are no studies that have evaluated whether or not the use of bedside ultrasound reduces infection, it is logical that fewer breaks in the skin surrounding the catheter site could only be beneficial. Certainly, bedside ultrasound does reduce mechanical complications. In a meta-analysis of 8 studies, the use of bedside ultrasound for the placement of CVCs significantly reduced mechanical complications compared to the standard landmark placement technique (22). Consideration of comfort, security, and maintenance of asepsis as well as patient-specific factors (e.g., preexisting catheters, anatomic deformity, bleeding diathesis), relative risk of mechanical complications (e.g., bleeding, pneumothorax), and the availability of bedside ultrasound should guide site selection.

Skin Antisepsis

In the United States, iodine- based disinfectants, particularly povidone iodine, has been the most widely used antiseptic for cleansing central catheter insertion sites. However, preparation of central venous and arterial sites with a 2% aqueous chlorhexidine gluconate has been shown to lower BSI rates compared to site preparation with 10% povidone-iodine or 70% alcohol (23). Chlorehexidine exhibits broad-spectrum antimicrobial activity on the skin surface after a single application, in contrast to iodine-based preparations. Commericially available products containing chlorhexidine have not been available in the United States until recently, when a 2% tincture of chlorhexidine preparation for skin antisepsis was approved by the FDA. Other preparations of chlorhexidine may not be as effective. Tincture of chlorhexidine gluconate 0.5% has not been shown to be more effective than 10% povidone iodine in adults. A prospective randomized study comparing 0.5% tincture of chlorhexidine gluconate to povidone iodine showed no difference in preventing CR-BSI or CVC colonization in adults (24). A 1% tincture of chlorhexidine preparation is available in Canada and Australia, but not yet in the United States. There are no published trials comparing a 1% chlorhexidine preparation to povidone-iodine. Based on the available data, a 2% chlorhexidine preparation should be used in patients to disinfect insertion sites, unless there is a contraindication to its use.

Antimicrobial/Antiseptic Impregnated Catheters

Certain antimicrobial or antisceptic impregnated or coated catheters have been shown to decrease the risk of catheter-related bloodstream infection in selected patient populations by up to 50%. Although these impregnated catheters cost more than the standard catheters, the potential decrease in hospital costs associated with treating CR-BSIs is not insignificant (25). Catheters coated with chlorhexidine/silver sulfadiazine only on the external luminal surface have been studied as a means to reduce CR-BSI. Two meta-analyses (26,27) demonstrated that the use of catheters coated on the external surface with chlorhexidine/silver sulfadiazine reduced the risk for CR-BI compared to standard non-coated catheters. The mean duration of catheter placement in one meta-analysis ranged between 5.1 and 11.2 days (28). The half-life of antimicrobial activity against S. epidermidis is 3 days in vitro for catheters coated with chlorhexidine/silver sulfadiazine, and the antimicrobial activity decreases over time (29). The benefit for the patients who receive these catheters will be realized within the first 14 days (28). This catheter is no no longer being marketed. Instead, a new second-generation catheter is now available with chlorhexidine coating both the internal and external luminal surfaces. The external surface has three times the amount of chlorhexidine and extended release of the surface bound antiseptics than that in the first generation catheters. Early studies indicate that prolonged anti-infective activity provides improved efficacy in preventing infections (30). Although rare, anaphylaxis has been reported with the use of these chlorhexidine/silver sulfadiazine catheters in Japan (31). As is the case with the use of any prophylactic antimicrobial drug, the risk of selecting organisms resistant to chlorhexidine/silver sulfadiazine is a concern; however this has not yet been demonstrated.

Chlorhexidine/silver sulfadiazine catheters are more expensive than standard catheters. However, one analysis has suggested that the use of chlorhexidine/silver sulfadiazine catheters should lead to a cost savings of $68 to $391 per catheter (32) in settings in which the risk of CR-BSI is high despite the adherence to other preventive strategies such as maximal barrier precautions and aseptic technique. These catheters may be cost effective when used in settings such as the intensive care unit. In a multicenter randomized trial, central venous catheters impregnated on the internal and external surfaces with minocycline/rifampin were associated with lower rates of CR-BSI when compared with the first generation chlorhexidine-silver sulfadiazine impregnated catheters (33). The beneficial effect began after day 6 of catheterization. None of the catheters were evaluated beyond 30 days. No minocycline/rifampin-resistant organisms were reported. However, based on in vitro data, there is concern that these impregnated catheters could increase the incidence of minocycline and rifampin resistance among important pathogens, especially staphylococci. The half-life of antimicrobial activity against *S. epidermidis* is 25 days with catheters coated with minocycline/rifampin, compared to 3 days for the first-generation catheters coated with chlorhexidine/silver sulfadiazine in vitro (29). *In vivo*, the duration of antimicrobial activity of the minocycline/rifampin catheter is longer than that of the first-generation chlorhexidine/silver sulfadiazine catheter (33). To date, no studies have been published using comparing the minocyline/rifampin catheter to the second-generation chlorhexidine/silver sulfadiazine catheter.

The decision to use chlorhexidine/silver sulfadiazine or minocycline/rifampin impregnated catheters should be based on the need to enhance prevention of CR-BSI after standard procedures (e.g. educating personnel, using full barrier precauntions and 2% chlorhexidine skin antisepsis) balanced against the concern for emergence of resistant pathogens and the cost of implementing this strategy.

Antiseptic/Antibiotic Ointments

Povidone-iodine ointment applied at the insertion site of hemodialysis catheters has been studied as a prophylactic intervention to reduce the incidence of catheter-related infections. One randomized study of hemodialysis catheters showed a reduction in the incidence of exit site infections, catheter-tip colonization and BSIs with the routine use of povidone-iodine ointment at the catheter insertion site compared to no ointment at the insertion site (34).

Several studies have evaluated the effectiveness of mupirocin ointment applied at the insertion sites of CVCs as a means to prevent CR-BSI (35-37). Although mupirocin reduced the risk for CR-BSI, mupirocin ointment has also been associated with mupirocin resistance (38, 39), and may adversely affect the integrity of polyurethane catheters (40, 41). Other antibiotic ointments applied to the catheter insertion site have also been studied and have yielded conflicting results (42-44). In addition, rates of catheter colonization with Candida species may be increased with the use of antibiotic ointments that have no fungicidal activity (42,44).

Nasal carriers of S. aureus have a higher risk for acquiring CR-BSI than do noncarriers (34,45). Mupirocin ointment has been used intranasally to decrease nasal carriage of S. aureus and lessen the risk for CR-BSI. However, resistance to mupirocin develops in both S. aureus and coagulase-negative staphylococci soon after routine use of mupirocin is instituted (38,39). The ecological impact of routine use of topical antimicrobial agents is not recommended because of the high likelihood of promoting antimicrobial resistance.

Antibiotic lock prophylaxis

The antibiotic lock technique is a novel method of using antibiotics as a local prophylaxis. Antibiotic lock prophylaxis has been attempted by filling the lumen of the catheter with an antibiotic solution and leaving the solution to dwell in the lumen of the catheter in order to prevent CR-BSI. Three studies have shown this to be useful in neutropenic patients with long-term catheters (46-48). In two of the studies, patients received heparin alone (10 U/ ml) or heparin plus 25 micrograms/ml of vancomycin. The third study compared vancomycin/ciprofloxacin/heparin to vancomycin/heparin to heparin alone. The rate of CR-BSI with vancomycin-susceptible organisms was significantly lower and the time to the first episode of bacteremia with vancomycin-susceptible organisms was significantly longer in patients receiving either vancomycin/ciprofloxacin/heparin or vancomycin/heparin compared to heparin alone.

Because most of these lock techniques have employed vancomycin as one component of the lock solution, there is concern over the potential effect of widespread use of prophylactic lock solutions on antimicrobial resistance. Therefore, these solutions are not routinely recommended. There use, howver, may be considered acceptable in selected patients who require long-term vascular access and who continue to experience infection despite adherence to standard infection control practices.

Catheter Site Dressing Regimens

Transparent semipermeable polyurethane dressings have become a popular means of dressing catheter insertion sites. Transparent dressings reliably secure the device, permit continuous visual inspection of the catheter site, permit patients to bathe and shower without saturating the dressing, and require less frequent changes than do standard gauze and tape dressings, thus saving personnel time. A meta-analysis of the largest and most rigorously controlled randomized trials has assessed these studies that compared the risk of catheter-related BSIs for groups using transparent dressings versus groups using gauze dressing (49). The risk for CR-BSIs did not differ between the groups. Therefore, the choice of dressing may be a matter of personal preference. If blood is oozing from the catheter insertion site, gauze dressing may be preferred.

An antiseptic dressing may be a better choice to reduce infection than the standard dressings. A chlorhexidine-impregnated sponge dressing (Biopatch) placed over the site of short-term arterial and CVCs reduced the risk of catheter colonization and CR-BSI in a multi-center study (50). There were no adverse systemic effects from using this device. Routine use of chlorhexidine–impregnated sponges may reduce the risk of CR-BSI in adult patients with short-term catheters, particularly with uncuffed CVCs, PICCs, and arterial catheters. However, chlorhexidine sponge dressings in neonates less than 7 days old or of gestational age less than 26 weeks have caused local reactions, precluding its use in this patient population (51).

Replacement of Catheters

The duration of catheterization has been linked to the risk of CR-BSI, particularly after 7 days. What has not been conclusively established is whether or not routine replacement of temporary CVCs at periodic intervals reduces the risk of CR-BSI. Although no studies have shown an advantage for routine catheter replacement at scheduled time intervals as a method to reduce CR-BSI, none have been sufficiently powered to demonstrate a difference. Two trials assessed changing the catheter every 7 days in comparison to changing catheters as needed (52,53). One study involved 112 surgical ICU patients needing central venous catheters, pulmonary artery catheters, or peripheral arterial catheters (52), while the other study involved only subclavian hemodialysis catheters (53). In both studies, there was no difference in CR-BSI in patients undergoing scheduled catheter replacement every 7 days compared to catheter replacement as needed. In the absence of data confirming a benefit, this practice is discouraged because of the potential mechanical complications associated with placing new catheters.

Scheduled guidewire exchanges of central catheters is another strategy that has been proposed to prevent CR-BSI. The results of a meta-analysis of 12 randomized controlled trials assessing central venous catheter management failed to prove any benefit for the reduction of CR-BSI by routine replacement of catheters by guidewire exchange compared to catheter replacement on an as-needed basis (54). Routine replacement of central venous catheters is not indicated for catheters that are functioning and have no evidence of local or systemic complications.

Catheter replacement over a guidewire has become an accepted technique for replacing a malfunctioning catheter or exchanging a pulmonary artery catheter for a central venous catheter when invasive monitoring no longer is needed. Catheter insertion over a guidewire is associated with less discomfort and a significantly lower rate of mechanical complications than are those percutaneously inserted at a new site (55) and provide an important means of preserving limited venous access in some difficult patients. However, maintaining asepstic technique during a guidewire exchange is difficult. Gloves and drapes are easily contaminated from manipulation of the old catheter.

Many studies have examined this practice on the risk of infection with conflicting results. The best study, however, showed an increased risk of CR-BSI with catheters replaced over a guidewire compared to catheters replaced at a new site as a routine replacement strategy. Replacement of temporary catheters over a guidewire in the setting of bacteremia is not an acceptable replacement strategy, since the source of infection is usually colonization of the skin tract from the insertion site to the vein (9,55).

CONCLUSIONS

CR-BSIs are not simply an acceptable consequence of central venous access and invasive monitoring. The false perception of invisible risk, the underestimation of individual responsibility, passive attitudes regarding the complexity of the process of care, and the financial constraints that contribute to understaffing play an important role in the failure to implement prevention strategies. Many CR-BSIs are preventable infections that need to be approached systematically at a multidisciplinary level that emphasizes patient safety and quality improvement. Taking advantage of new technology such as chlorehexidine for skin antisepsis, and antiseptic or antibiotic impregnated catheters, chlorehexidine impregnated dressings, may be a useful means of reducing catheter-related infections. What is not known is whether each technology contributes additively, synergistically, or not at all. In other words, does one need all of these strategies, or will one of them suffice? In the absence of data, it seems logical to utilize these strategies in a systematic way that incorporates performance measures to evaluate the impact that each intervention has in a given ICU setting.

REFERENCES

1. Wenzel RP, Edmond MB. The evolving technology of venous access. N Engl J Med 1999;340:48-50.
2. Dimick JB, Pelz RK, Consunji R, Swoboda SM, Hendrix CW, Lipsett PA. Increased resource use associated with catheter-related bloodstream infection in the surgical intensive care unit. Arch Surg 2001;136:229-34.
3. Rello J, Ochagavia A, Sabanes E, Roque M, Mariscal D, Reynaga E, Valles J. Evaluation of outcome of intravenous catheter-related infections in critically ill patients. Am J Respir Crit Care Med 2000;162:1027-30.
4. Heiselman D. Nosocomial bloodstream infections in the critically ill. JAMA 1994;272:1819-20.
5. Centers for Disease Control. National Nosocomial Infections Surveillance (NNIS) System report, data summary from January 1992-June 2001, issued August 2001. Am J Infect Control 2001;in press.
6. Centers for Disease Control. National Nosocomial Infections Surveillance (NNIS) System report, data summary from January 1990-May 1999, issued June 1999. Am J Infect Control 1999;27:520-532.
7. Schaberg DR, Culver DH, Gaynes RP. Major trends in the microbial etiology of nosocomial infection. Am J Med 1991;91:72S-75S.
8. Maki DG, Weise CE, Sarafin HW. A semiquantitative culture method for identifying intravenous-catheter-related infection. N Engl J Med 1977;296:1305-9.
9. Mermel LA, McCormick RD, Springman SR, Maki DG. The pathogenesis and epidemiology of catheter-related infection with pulmonary artery Swan-Ganz catheters: a prospective study utilizing molecular subtyping. Am J Med 1991;91:197S-205S.
10. Raad I, Costerton W, Sabharwal U, Sacilowski M, Anaissie E, Bodey GP. Ultrastructural analysis of indwelling vascular catheters: a quantitative relationship between luminal colonization and duration of placement. J Infect Dis 1993;168:400-7.
11. Locci R, Peters G, Pulverer G. Microbial colonization of prosthetic devices. IV. Scanning electron microscopy of intravenous catheters invaded by yeasts. Zentralbl Bakteriol Mikrobiol Hyg [B] 1981;173:419-24.
12. Locci R, Peters G, Pulverer G. Microbial colonization of prosthetic devices. I. Microtopographical characteristics of intravenous catheters as detected by scanning electron microscopy. Zentralbl Bakteriol Mikrobiol Hyg [B] 1981;173:285-92.
13. Nachnani GH, Lessin LS, Motomiya T, Jensen WN, Bodey GP. Scanning electron microscopy of thrombogenesis on vascular catheter surfaces Ultrastructural analysis of indwelling vascular catheters: a quantitative relationship between luminal colonization and duration of placement. N Engl J Med 1972;286:139-40.
14. Stillman RM, Soliman F, Garcia L, Sawyer PN. Etiology of catheter-associated sepsis. Correlation with thrombogenicity. Arch Surg 1977;112:1497-9.

15.   Eggimann P, Harbarth S, Constantin MN, Touveneau S, Chevrolet JC, Pittet D. Impact of a prevention strategy targeted at vascular-access care on incidence of infections acquired in intensive care. Lancet 2000;355:1864-8.

16.   Murphy LM, Lipman TO. Central venous catheter care in parenteral nutrition: a review. JPEN J Parenter Enteral Nutr 1987;11:190-201.

17.   Sanders RA, Sheldon GF. Septic complications of total parenteral nutrition. A five year experience. Am J Surg 1976;132:214-20.

18.   Sherertz RJ, Ely EW, Westbrook DM, Gledhill KS, Streed SA, Kiger B, Flynn L, Hayes S, Strong S, Cruz J, Bowton DL, Hulgan T, Haponik EF. Education of physicians-in-training can decrease the risk for vascular catheter infection. Ann Intern Med 2000;132:641-8

19.   Nehme AE. Nutritional support of the hospitalized patient. The team concept. JAMA 1980;243:1906-8.

20.   Soifer NE, Borzak S, Edlin BR, Weinstein RA. Prevention of peripheral venous catheter complications with an intravenous therapy team: a randomized controlled trial. Arch Intern Med 1998;158:473-7.

21.   Tomford JW, Hershey CO. The i.v. therapy team: impact on patient care and costs of hospitalization. NITA 1985;8:387-9.

22.   Randolph AG, Cook DJ, Gonzales CA, Pribble CG. Ultrasound guidance for placement of central venous catheters: a meta-analysis of the literature. Crit Care Med 1996;24:2053-8.

23.   Maki DG, Ringer M, Alvarado CJ. Prospective randomised trial of povidone-iodine, alcohol, and chlorhexidine for prevention of infection associated with central venous and arterial catheters. Lancet 1991;338:339-43.

24.   Humar A, Ostromecki A, Direnfeld J, Marshall JC, Lazar N, Houston PC, Boiteau P, Conly JM. Prospective randomized trial of 10% povidone-iodine versus 0.5% tincture of chlorhexidine as cutaneous antisepsis for prevention of central venous catheter infection. Clin Infect Dis 2000;31:1001-7.

25.   Raad I, Darouiche R, Dupuis J, Abi-Said D, Gabrielli A, Hachem R, Wall M, Harris R, Jones J, Buzaid A, Robertson C, Shenaq S, Curling P, Burke T, Ericsson C. Central venous catheters coated with minocycline and rifampin for the prevention of catheter-related colonization and bloodstream infections. A randomized, double-blind trial. The Texas Medical Center Catheter Study Group. Ann Intern Med 1997;127:267-74.

26.   Mermel LA. Prevention of intravascular catheter-related infections. Ann Intern Med 2000;132:391-402.

27.   Veenstra DL, Saint S, Saha S, Lumley T, Sullivan SD. Efficacy of antiseptic-impregnated central venous catheters in preventing catheter-related bloodstream infection: a meta-analysis. JAMA 1999;281:261-7.

28.   Maki DG, Stolz SM, Wheeler S, Mermel LA. Prevention of central venous catheter-related bloodstream infection by use of an antiseptic-impregnated catheter. A randomized, controlled trial. Ann Intern Med 1997;127:257-66.

29.   Raad I, Darouiche R, Hachem R, Mansouri M, Bodey GP. The broad-spectrum activity and efficacy of catheters coated with minocycline and rifampin. J Infect Dis 1996;173:418-24

30.   Bassetti S, Hu J, D'Agostino RB, Jr., Sherertz RJ. Prolonged antimicrobial activity of a
       catheter containing chlorhexidine-silver sulfadiazine extends protection against
       catheter infections *in vivo*. Antimicrob Agents Chemother 2001;45:1535-8.
31.   Oda T, Hamasaki J, Kanda N, Mikami K. Anaphylactic shock induced by an
       antiseptic-coated central venous [correction of nervous] catheter. Anesthesiology
       1997;87:1242-4
32.   Veenstra DL, Saint S, Sullivan SD. Cost-effectiveness of antiseptic-impregnated
       central venous catheters for the prevention of catheter-related bloodstream infection.
       JAMA 1999;282:554-60.
33.   Darouiche RO, Raad, II, Heard SO, Thornby JI, Wenker OC, Gabrielli A, Berg J,
       Khardori N, Hanna H, Hachem R, Harris RL, Mayhall G. A comparison of two
       antimicrobial-impregnated central venous catheters. Catheter Study Group. N Engl J
       Med 1999;340:1-8.
34. ' Levin A, Mason AJ, Jindal KK, Fong IW, Goldstein MB. Prevention of hemodialysis
       subclavian vein catheter infections by topical povidone-iodine. Kidney Int
       1991;40:934-8.
35.   Casewell MW. The nose: an underestimated source of Staphylococcus aureus causing
       wound infection. J Hosp Infect 1998;40:S3-11.
36.   Hill RL, Fisher AP, Ware RJ, Wilson S, Casewell MW. Mupirocin for the reduction of
       colonization of internal jugular cannulae--a randomized controlled trial. J Hosp Infect
       1990;15:311-21.
37.   Sesso R, Barbosa D, Leme IL, Sader H, Canziani ME, Manfredi S, Draibe S, Pignatari
       AC. Staphylococcus aureus prophylaxis in hemodialysis patients using central venous
       catheter: effect of mupirocin ointment. J Am Soc Nephrol 1998;9:1085-92.
38.   Miller MA, Dascal A, Portnoy J, Mendelson J. Development of mupirocin resistance
       among methicillin-resistant Staphylococcus aureus after widespread use of nasal
       mupirocin ointment. Infect Control Hosp Epidemiol 1996;17:811-3.
39.   Zakrzewska-Bode A, Muytjens HL, Liem KD, Hoogkamp-Korstanje JA. Mupirocin
       resistance in coagulase-negative staphylococci, after topical prophylaxis for the
       reduction of colonization of central venous catheters. J Hosp Infect 1995;31:189-93.
40.   Rao SP, Oreopoulos DG. Unusual complications of a polyurethane PD catheter. Perit
       Dial Int 1997;17:410-2.
41.   Riu S, Ruiz CG, Martinez-Vea A, Peralta C, Oliver JA. Spontaneous rupture of
       polyurethane peritoneal catheter. A possible deleterious effect of mupirocin ointment.
       Nephrol Dial Transplant 1998;13:1870-1.
42.   Maki DG, Band JD. A comparative study of polyantibiotic and iodophor ointments in
       prevention of vascular catheter-related infection. Am J Med 1981;70:739-44.
43.   Norden CW. Application of antibiotic ointment to the site of venous catheterization--a
       controlled trial. J Infect Dis 1969;120:611-5.
44.   Zinner SH, Denny-Brown BC, Braun P, Burke JP, Toala P, Kass EH. Risk of infection
       with intravenous indwelling catheters: effect of application of antibiotic ointment. J
       Infect Dis 1969;120:616-9.
45.   von Eiff C, Becker K, Machka K, Stammer H, Peters G. Nasal Carriage as a Source of
       Staphylococcus aureus Bacteremia. N Engl J Med 2001;344:11-16.

46.   Carratala J, Niubo J, Fernandez-Sevilla A, Juve E, Castellsague X, Berlanga J, Linares J, Gudiol F. Randomized, double-blind trial of an antibiotic-lock technique for prevention of gram-positive central venous catheter-related infection in neutropenic patients with cancer. Antimicrob Agents Chemother 1999;43:2200-4.

47.   Henrickson KJ, Axtell RA, Hoover SM, Kuhn SM, Pritchett J, Kehl SC, Klein JP. Prevention of central venous catheter-related infections and thrombotic events in immunocompromised children by the use of vancomycin/ciprofloxacin/heparin flush solution: A randomized, multicenter, double-blind trial. J Clin Oncol 2000;18:1269-78.

48.   Schwartz C, Henrickson KJ, Roghmann K, Powell K. Prevention of bacteremia attributed to luminal colonization of tunneled central venous catheters with vancomycin-susceptible organisms. J Clin Oncol 1990;8:1591-7.

49.   Hoffmann KK, Weber DJ, Samsa GP, Rutala WA. Transparent polyurethane film as an intravenous catheter dressing. A meta-analysis of the infection risks. JAMA 1992;267:2072-6.

50.   Maki DG, Mermel LA, Klugar D, Narans L, Knasinski V, Parenteau S, Covington P. The efficacy of a chlorhexidine impregnated sponge (Biopatch)for the prevention of intravascular catheter-related infection- a prospective randomized controlled multicenter study. In: ICAAC. Toronto, Ontario, Canada:, 2000.

51.   Garland JS, Alex CP, Mueller CD, Cisler-Kahill LA. Local reactions to a chlorhexidine gluconate-impregnated antimicrobial dressing in very low birth weight infants. Pediatr Infect Dis J 1996;15:912-4.

52.   Eyer S, Brummitt C, Crossley K, Siegel R, Cerra F. Catheter-related sepsis: prospective, randomized study of three methods of long-term catheter maintenance. Crit Care Med 1990;18:1073-9.

53.   Uldall PR, Merchant N, Woods F, Yarworski U, Vas S. Changing subclavian haemodialysis cannulas to reduce infection. Lancet 1981;1:1373.

54.   Cook D, Randolph A, Kernerman P, Cupido C, King D, Soukup C, Brun-Buisson C. Central venous catheter replacement strategies: a systematic review of the literature. Crit Care Med 1997;25:1417-24.

55.   Cobb DK, High KP, Sawyer RG, Sable CA, Adams RB, Lindley DA, Pruett TL, Schwenzer KJ, Farr BM. A controlled trial of scheduled replacement of central venous and pulmonary-artery catheters. N Engl J Med 1992;327:1062-8.

INDEX

Acinetobacter calcoaceticus, 161
Acinetobacter species, 104, 105
Acridine-orange leukocyte cytospin test, 53, 70–71
Adherence, bacterial, mechanisms of, 33–34
Aerobic gram-negative bacilli, 104–105
Aerobic gram-positive bacilli, 104
Aeruginosa pseudomonas, 104
AIDS, risk of infection with, 11
Aminoglycoside, 122
Amphotericin, 122
Ampicillin aminoglycoside, 122
Ampicillin-resistant infection, antimicrobial therapy, 122
Ampicillin-sensitive infection, antimicrobial therapy, 122
Antibiotic catheter coating
    antiseptic catheter coating, rates of infection, compared, 12
    rates of infection, 5, 12
Antibiotic lock therapy, 107, 117–118, 166
Anti-infective cream, topical, risk of infection, 11
Antimicrobial catheter coating, 153–154, 163–165
Antimicrobial therapy, 117–122
    antibiotic lock therapy, 107, 117–118, 166
    systemic, 118–119
Antisepsis
    cutaneous, 16–19
    skin, 163
Antiseptic catheter coating, 153–154, 163–165
    antibiotic catheter coating, rates of infection, compared, 12
    rates of infection with, 12
Antiseptic/antibiotic ointments, 165
Antiseptic-impregnated catheters, rates of infection caused by, 5
Antithrombotic prophylaxis, in prevention strategy, 150–151
AOLC test. *See* Acridine-orange leukocyte cytospin test
APACHE III score, high, risk of infection with, 11
Arterial catheters
    for hemodynamic monitoring, rates of infection caused by, 6
    risk factors for catheter-related infection with, 13
Arterial puncture, educational presentation, 131

Asinetobacter species, 101

Bacillus species, 32, 36, 101, 104
Bacteremia
    as indication for removal of non-tunneled central venous catheter, 116
    in intensive care units, catheter-related, case control studies, 83
Bacterial adherence, mechanisms of, 33–34
Barriers, 14, 148
    garments, educational presentation, 131
    sterile, maximal, risk of infection with, 11
Bedside ultrasound, as infection preventive strategy, 162–163
Birth weight, low, risk of infection with, 11
Blood cultures, quantitative, 67
Blood sampling
    educational information on, 132
    rates of infection with, 12
Bloodstream infection. *See* specific infection, organism

Candida albicans, 105, 115, 116, 119, 122, 161
Candida glabarata, 105, 161
Candida krusei, 105, 161
Candida parapsilosis, 101, 105
Candida species, 36, 101, 105, 106, 108, 116, 117, 119, 122, 123, 151
    antimicrobial therapy, 105–106
Candida tropicalis, 105
Cardiac output, educational information on, 132
Catheter characteristics, rates of infection according to, 12
Catheter colonization, defined, 3
Catheter exchange over guidewire, 15–16
Catheter hub, culture of, 65–66
Catheter removal, 114–115
Catheter salvage strategies, 106–107
    antibiotic lock therapy, 107
Catheter staining, in diagnosis of catheter infection, 46
Catheter types, rates of infection caused, 5
Catheter-related infection in critically ill
    diagnosis, 41–58, 59–76
    education for prevention, 127–138, 139–145. *See also* Prevention
    epidemiology of, 1–22, 23–40
    impact of, 77–86
    management, 99–112, 113–126
    pathogenesis, 23–40

prevention, 10–16, 127–138, 139–145,
    147–158, 159–172, 167–168
treatment, 99–112, 113–126
Catheter-tip culture techniques in diagnosis, 61
    qualitative broth culture, 61
    semi-quantitative catheter-tip culture, 62
Cefotaxime, 122
Ceftazidime, 122
Central, peripheral quantitative blood cultures,
    paired, 67–69
Central blood culture methods
    quantitative, 67
Central vein infection, defined, 4
Central venous catheter, infection, 114
    exchanged over guidewire, with significant
        colonization, 116
    rates of, 5
Chlorhexidine vs. povidone iodine, risk of
    infection, compared, 11
Chlorhexidine-impregnated dressing, risk of
    infection, 12
Ciprofloxacin, 122
Coagulase-negative staphylococcus, 101–104,
    123
    antimicrobial therapy, 102, 122
Coexistence of other intravascular devices, risk
    of infection with, 11
Collagen-silver impregnated cuffs, 152–153
Colonization, infection, distinguished, 100
Complicated infection, management of,
    107–108
    endocarditis, 108
    septic thrombophlebitis, 108
Contamination-resistant hub, rates of infection
    with, 12
Continuing education. See Education
Corynebacterium jeikeium, 104
Corynebacterium species, 32
Cost effectiveness, of education, for
    prevention, 134
Costs, of catheter-related infection, 91, 81084
Critically ill, catheter-related infection in
    education for prevention, 127–138, 139–145.
        See also Prevention
    prevention, 10–16, 127–138, 139–145,
        147–158, 159–172, 167–168
Cuffed catheters, rates of infection caused by, 5
Cuffed central venous catheters, rates of
    infection caused by, 5
Cutaneous antiseptics, 11, 16–19
Cutdowns, rates of infection caused by, 5

Dalfopristin, 121, 122
Defatting insertion site, risk of infection, 11
Diagnosis of catheter-related infections, 41–58,
    59–76
Difficult insertion, risk of infection with, 11
Direct examination, rapid diagnosis of infection
    by, 70–71

acridine-orange leukocyte cytospin test,
    70–71
gram staining of blood drawn from catheter,
    71
Dressings, 151–152
    educational information on, 132
    regimens, at catheter site, in infection
        preventive strategy, 166–167
    site, 18
    transparent, use of, 132
    types, risks of infection, compared, 12
Dry gauze, use of, 132
Education, in catheter-related infection
        prevention, 127–138, 139–145, 162
    cost effectiveness, 134
    educational programs, 128–129
    general measures, 128
    Geneva experience, 129–130
    guidelines, 128
    impact of, 36, 135
    North Carolina experience, 129
    St. Louis experience, 134
    training, 128–129

Endocarditis, 108
Endoluminal brush, in infection diagnosis,
    51–52
Endoluminal contamination, routes of, 29–31.
    See also specific route
Enterobacter agglomerans, 105
Enterobacteriaceae, extraluminal
    contamination, 28
Enterococcus species, 28
Entry site, culture of, 65–66
Epidemiology, catheter-related infection, 1–22,
    160–161
    catheter exchange over guidewire, 15–16
    cutaneous antisepsis, 16–19
    dressings, site, 18
    experience of inserter, 13
    heavy colonization, insertion site, 16–19
    intravenous therapy teams, 13–14
    microbiology, 6–10
    pathogenesis, 6
    prevention, 10–16
    prolonged catheter placement, 18–19
    risk factors, 10–16
    site of insertion, 14–15
    sterile barrier precautions, 14
    training of inserter, 13
Erythema, in skin site sepsis, as indication for
        removal of non-tunneled central venous
        catheter, 116
Escherichia coli, 36, 104, 122
Exit-site infection. See also specific infection,
        organism
    overview, 3
Extended hospitalization, risk of infection with,
    11

Extraluminal contamination, routes of, 27–28.
    See also specific route
Exudate, purulent, in skin site sepsis, as
    indication for removal of non-tunneled
    central venous catheter, 116

Femoral vein insertion, risk of infection, 11
Filamentous fungi, 106
Flucloxacillin, 122
Fluconazole, 122
Fungi, antimicrobial therapy, 122
Fusarium species, 101, 106

Gastrointestinal disease, risk of infection with,
    11
Gauze dressing
    dry, use of, 132
    vs. transparent dressing, risk of infection
        with, compared, 12
Gender, risk of infection with, 11
Gram staining of blood drawn from catheter, 71
Gram-negative bacilli, antimicrobial therapy,
    122
Gram-positive cocci, antimicrobial therapy, 122
Guidewire
    catheter exchange, 15–16
    diagnosis of infection, 72
    central venous catheter exchanged over, with
        significant colonization, 116
    insertion in old site over, risk of infection
        with, 11

Hand disinfection, educational information on,
    132
Hand hygiene procedures, educational
    presentation, 131
Hansenula anomala, 101
Heavy colonization, insertion site, 16–19
Hemodialysis catheters, rates of infection
    caused by, 5
Hemodynamic monitoring, arterial catheters,
    rates of infection caused by, 6
Heparin-bonded catheters, rates of infection
    caused by, 5
Heparinization, rates of infection with, 12
High APACHE III score, risk of infection with,
    11
House-staff, insertion of by, risk of infection
    with, 11
Hubs
    number of, as risk factor, 35
    originated, routes of, 29–31. *See also*
        specific route

Impact of catheter-related infection, 77–86,
    87–98. *See also* Costs; Morbidity;
    Mortality
Implementation of ICU catheter infection
    prevention program, 154

Infection, catheter-related, in critically ill
    diagnosis, 41–58, 59–76
    education for prevention, 127–138, 139–145.
        See also Prevention
    epidemiology of, 1–22, 23–40
    impact of, 77–86
    management, 99–112, 113–126
    pathogenesis, 23–40
    prevention, 10–16, 127–138, 139–145,
        147–158, 159–172, 167–168
    treatment, 99–112, 113–126
Infusate-related infection, 35–36
    defined, 3
Inserter
    experience of, 13
    training of, 13
Insertion of catheter, prevention of infection,
    131, 148–150
    cutaneous antisepsis, 148–149
    site, 149–150
    sterile barrier precautions, 148
    tunneling, 150
Insertion site, 14–15
    features of, risk of infection with, 11
    infection diagnosis, 47
    in internal jugular, risk of infection with, 13
Intensive care unit, prevention strategies in, 11,
        83, 147–158, 159–172
    antibiotic lock therapy, 166
    antimicrobial/antiseptic-impregnated
        catheters, 163–165
    antiseptic/antibiotic ointments, 163 165
    catheter, 151–154
        antimicrobial-coated/impregnated
            catheters, 153–154
        collagen-silver impregnated cuffs,
            152–153
        cutaneous antisepsis, 148–149
        dressing, 151–152
        insertion, 148–150
        site, 149–150
        replacement, 151
        sterile barrier precautions, 148
        tubing, 151–152
        tunneling, 150
    continuing education, 162
    dressing regimens, at catheter site, 166–167
    epidemiology, 160–161
    implementation of, 154
    mechanisms of infection, 148
    microbiology, 160–161
    novel strategies, 162–168
    pathogenesis, 161
    prophylactic treatments, 150–151
        antithrombotic prophylaxis, 150–151
    quality assurance, 162
    replacement, catheter, 167–168
    ultrasound, bedside, 162–163

Intensive care unit nurses, central venous
    catheter insertion/maintenance, content of
    education program for, 133
Internal jugular vein, insertion site in, risk of
    infection with, 11, 13
Intravenous therapy teams, 13–14

Jugular, internal, insertion site in, risk of
    infection with, 11, 13
Klebsiella pneumoniae, 104, 161
Klebsiella species, 122

Length of prolonged catheter placement, 18–19
Linezolid, 122
Local infections, diagnosis of, 60
Lock therapy, antibiotic, 107, 117–118, 166
    in catheter-related infection prevention, 166
Low birth weight, risk of infection with, 11
Lumbar puncture, educational presentation, 131

Malassezia furfur, 101, 106
Management, 113–126
    antimicrobial therapy, 117–122
        antibiotic lock therapy, 117–118
        systemic antimicrobial therapy, 118–119
    catheter removal, 114–115
    central venous catheter infections, 114
    non-tunneled central venous catheter,
        115–116
    quinupristin-dalfopristin, 121
    tunneled central venous catheter, 116–117
    vancomycin-resistant staphylococci, 121
Mechanical ventilation, risk of infection with,
    11
Mechanisms of bacterial adherence, 33–34
Metastatic infection, as indication for removal
    of non-tunneled central venous catheter,
    116
Methicillin-resistant infection, antimicrobial
    therapy, 122
Methicillin-sensitive infection, antimicrobial
    therapy, 122
Microbiology of catheter-related infection,
    6–10, 160–161
Microorganism types, interaction with catheter
    material, 31–34. See also specific
    microorganism
Microorganism-directed therapy, 101–106
    aerobic gram-negative bacilli, 104–105
        antimicrobial therapy, 105
    aerobic gram-positive bacilli, 104
    Candida species, 105
        antimicrobial therapy, 105–106
    coagulase-negative staphylococcus, 101–104
        antimicrobial therapy, 102
    filamentous fungi, 106
    rapidly growing mycobacteria, 106
    Staphylococcus aureus, 102
        antimicrobial therapy, 103–104

Morbidity, with catheter-related infection,
    81084
Mortality, from catheter-related infection,
    78–81, 91, 93
Multi-lumen catheter, single-lumen catheter,
    rates of infection, compared, 12
Mupirocin, risk of infection, 11
Mycobacterium abscessus, 101, 106
Mycobacterium chelonei, 101
Mycobacterium fortuitum, 101

National Nosocomial Infections Surveillance
    System, 42
Neutropenia, risk of infection with, 11
Noncuffed catheters
    nonmedicated central venous catheters, rates
        of infection caused by, 5
    pulmonary-artery catheters, risk factors for
        catheter-related infection with, 13
    rates of infection caused by, 5
Non-tunneled central venous catheter, 115–116
Novel strategies for prevention of
    catheter-related infection, 162–168
Number of hubs, as risk factor, 35
Nurses, intensive care unit, central venous
    catheter insertion/maintenance, content of
    education program for, 133

Ointments, antiseptic/antibiotic, in
    catheter-related infection prevention, 165

Paired peripheral non-quantitative blood
    cultures, central, 69–70
Paired quantitative central blood culture
    methods, 68
Parenteral nutrition, 35–36
    infusate-related infection and, 35–36
    rates of infection with, 12
Pathogenesis, catheter-related infection, 6,
    59–60, 161
Patient population, critically ill, catheter-related
    infection in
    diagnosis, 41–58, 59–76
    education for prevention, 127–138, 139–145.
        See also Prevention
    epidemiology of, 1–22, 23–40
    impact of, 77–86
    management, 99–112, 113–126
    pathogenesis, 23–40
    prevention, 10–16, 127–138, 139–145,
        147–158, 159–172, 167–168
    treatment, 99–112, 113–126
Peripheral, central quantitative blood cultures,
    paired, 67–69
Peripheral venous catheter insertion,
    educational presentation, 131
Phlebitis, defined, 3
Phlebotomy, educational presentation, 131
Plastic catheters, rates of infection caused by, 5

Pocket infection, defined, 3
Povidone iodine
   risk of infection, 11
   vs. chlorhexidine, risk of infection,
      compared, 11
Povidone-iodine, for skin preparation, 131
Prevention strategies in intensive care unit,
      147–158, 159–172
   antibiotic lock therapy, 166
   antimicrobial/antiseptic-impregnated
      catheters, 163–165
   antisepsis, skin, 163
   antiseptic/antibiotic ointments, 165
   antithrombotic prophylaxis, 150–151
   catheter, 151–154
      antimicrobial-coated/impregnated
         catheters, 153–154
      collagen-silver impregnated cuffs,
         152–153
      dressing, 151–152
      insertion, 148–150
         cutaneous antisepsis, 148–149
         site, 149–150
         sterile barrier precautions, 148
      replacement, 151
      tubing, 151–152
      tunneling, 150
   continuing education, 162
   dressing regimens, at catheter site, 166–167
   epidemiology, 160–161
   implementation of, 154
   mechanisms of infection, 148
   microbiology, 160–161
   novel strategies, 162–168
   pathogenesis, 161
   prophylactic treatments, 150–151
   quality assurance, 162
   replacement, catheter, 167–168
   ultrasound, bedside, 162–163
Prolonged catheter placement, 18–19
Prophylactic treatments, 150–151. See also
      under specific treatment
   antithrombotic prophylaxis, 150–151
Pseudomonas aeruginosa, 78, 101, 106, 115,
      118, 119, 122, 161
Pseudomonas species, 28, 104, 105
Pulmonary-artery catheters, rates of infection
      caused by, 5
Pulsed-field gel electrophoresis, of
      pathogenesis, 8
Purulent exudate, in skin site sepsis, as
      indication for removal of non-tunneled
      central venous catheter, 116

Quality assurance, for catheter-related infection
      prevention, 162
Quantitative blood cultures, 67
   in infection diagnosis, 47

Quantitative catheter-tip culture techniques in
      diagnosis, 63
   catheter culture techniques, compared, 64
   in pulmonary artery catheters, 64
Quantitative skin cultures, in infection
      diagnosis, 48
Quinupristin/dalfopristin, 121, 122

Rapidly growing mycobacteria, 106
Rates of bloodstream infection, by type of
      device, 5. See also Routes of catheter
      contamination
Registered ICU nurses, central venous catheter
      insertion/maintenance, content of
      education program for, 133
Removal of catheter, diagnosis of infection
      without, 65–70
Replacement of catheter, in infection
      preventive strategy, 167–168
Rhodotorula species, 101
Risk factors for catheter infections, 10–16,
      34–36
Routes of catheter contamination, 27–31

Semiquantitative cultures, in diagnosis of
      catheter infection, 43–44
Septic thrombophlebitis, 108. See also specific
      infection, organism
   overview, 4
Septicemia, as indication for removal of
      non-tunneled central venous catheter, 116
Serological tests, in infection diagnosis, 53
Silver-impregnated cuff, 5, 12
Single-lumen catheter, multi-lumen catheter,
      rates of infection, compared, 12
Skin exit-site culture method, 66
Skin exit-site originated contamination routes,
      27–28. See also specific route
Skin preparation, educational information on,
      132
Sources of intravenous device-related
      bloodstream infection, overview, 6
Spectrum of catheter-related infection:, 3
Standard noncuffed, nonmedicated central
      venous catheters, rates of infection caused
      by, 5
Staphylococcus hominis, 32
Staphylococcus aureus, 7, 26, 28, 29, 32, 35,
      53, 81, 87, 88, 90, 91, 93, 95, 99, 100,
      101, 102, 103–104, 106, 108, 115, 116,
      117, 118, 119, 120, 122–123, 123, 160
Staphylococcus epidermidis, 32, 33, 35, 53,
      100, 101, 116, 164
Staphylococcus haemolyticus, 32, 119
Staphylococcus hominis, 32
Steel needles, rates of infection caused by, 5
Stenotrophomonas maltophilia, 36, 101, 104,
      105
Sterile barriers, 14

for insertion, educational presentation, 131
  maximal, risk of infection with, 11
Student, insertion of by, risk of infection with,
  11
Subclavian vein insertion, risk of infection, 11
Subcutaneous central ports, rates of infection
  caused by, 5
Surgical cutdown, insertion using, 13
Surgical procedure, risk of infection with, 11
Surgically-implanted long-term central venous
  devices
  infection, 10
  percutaneously-inserted short-term
    noncuffed intravenous devices,
    infection, 10

Therapy strategies
  antimicrobial therapy, 117–122
    antibiotic lock therapy, 117–118
    systemic antimicrobial therapy, 118–119
  microorganism-directed therapy
    aerobic gram-negative bacilli,
      antimicrobial therapy, 105
    Candida species, antimicrobial therapy,
      105–106
    coagulase-negative staphylococcus,
      antimicrobial therapy, 102
    Staphylococcus aureus, antimicrobial
      therapy, 103–104
Thrombophlebitis, septic, 108. See also specific
  infection, organism
  overview, 4
Thrombosis, septic. See also specific infection,
  organism
  overview, 4
Topical anti-infective cream, risk of infection,
  11
Transparent dressing
  use of, 132
  vs. gauze dressing, risk of infection with,
    compared, 12
Transplantation, risk of infection with, 11
Treatment strategies, 99–112, 113–126
  bloodstream invasion, etiology of, 100–101
  catheter removal, 114–115

catheter salvage strategies, 106–107
  antibiotic lock therapy, 107
central venous catheter infections, 114
colonization, infection, distinguished, 100
complicated infection, 107–108
  endocarditis, 108
  septic thrombophlebitis, 108
microorganism-directed therapy, 101–106
  aerobic gram-negative bacilli, 104–105
  aerobic gram-positive bacilli, 104
  Candida species, 105
  coagulase-negative staphylococcus,
    101–104
    filamentous fungi, 106
    rapidly growing mycobacteria, 106
    Staphylococcus aureus, 102
non-tunneled central venous catheter,
  115–116
quinupristin-dalfopristin, 121
tunneled central venous catheter, 116–117
vancomycin-resistant staphylococci, 121
Trichosporon beigelii, 101
Trichosporon species, 101, 106
Tubing, in infection prevention, 151–152
Tunnel infection, defined, 3
Tunneled catheter, 5, 116–117
  noncuffed, rates of infection caused by, 5
Tunneling non-cuffed central venous catheter,
  maximal, 11

Ultrasound, bedside, in infection preventive
  strategy, 162–163
Under staffing, impact of, 36
Urinary catheter insertion, educational
  presentation, 131

Vancomycin, 122
Vancomycin aminoglycoside, 122
Vancomycin resistance, 121, 122
Vancomycin-sensitive infection, antimicrobial
  therapy, 122

Washing of hands, educational information on,
  132